ACKNOWLEGMENTS

The development of this document was supported by individuals listed below, who each brought unique experience and insights to the development.

AUTHORS:

Dr. Homayoon Dezfuli (Project Manager)	NASA Headquarters
Dr. Michael Stamatelatos	NASA Headquarters
Mr. Gaspare Maggio	Information Systems Laboratories
Mr. Christopher Everett	Information Systems Laboratories
Dr. Robert Youngblood	Idaho National Laboratory

CONTRIBUTING AUTHORS:

Dr. Peter Rutledge	Quality Assurance & Risk Management Services
Dr. Allan Benjamin	ARES Corporation
Dr. Rodney Williams	Information Systems Laboratories
Dr. Curtis Smith	Idaho National Laboratory
Dr. Sergio Guarro	Aerospace Corporation

REVIEWERS:

This development benefited from review comments provided on the initial draft by many individuals. The authors wish to specifically thank the following individuals:

Dr. Robert Abelson	NASA Jet Propulsion Laboratory
Dr. Timothy Barth	NASA Kennedy Space Center
Mr. Alfredo Colon	NASA Headquarters
Mr. Chester Everline	NASA Jet Propulsion Laboratory
Mr. Louis Fussel	Futron Corporation
Dr. Frank Groen	NASA Headquarters
Mr. David Lengyel	NASA Headquarters
Dr. Robert Mulvihill	Quality Assurance & Risk Management Services
Mr. William Powell	NASA Marshall Space Flight Center
Mr. David Pye	Perot Systems
Dr. James Rose	NASA Jet Propulsion Laboratory
Dr. Fayssal Safie	NASA Marshall Space Flight Center
Dr. Nathan Siu	U.S. Nuclear Regulatory Commission
Mr. Clayton Smith	Johns Hopkins Applied Physics Laboratory

Published by Books Express Publishing
Copyright © Books Express, 2012
ISBN 978-1-78266-141-2

Books Express publications are available from all good retail and online booksellers. For publishing proposals and direct ordering please contact us at: info@books-express.com

TABLE OF CONTENTS

LIST OF FIGURES

LIST OF TABLES

RISK-INFORMED DECISION MAKING
In the Context of NASA Risk Management

Preface

Risk management (RM) is an integral aspect of virtually every challenging human endeavor, but well-defined RM processes have only recently begun to be developed and implemented as an integral part of systems engineering at NASA, given the complex concepts that RM encapsulates and the many forms it can take. However, few will disagree that effective risk management is critical to program and project success.

Recent NASA RM processes have been based on Continuous Risk Management (CRM), which stresses the management of risk during implementation. In December of 2008, NASA issued NPR 8000.4A [1], which introduced Risk-Informed Decision Making (RIDM) as a complementary process to CRM that is concerned with analysis of important and/or direction-setting decisions. Before, RM was considered equivalent to CRM; now, RM is defined as comprising both CRM and RIDM.

This handbook addresses the RIDM component of RM. This is an essential part of RM since the decisions made during the course of a program ultimately "burn-in" the risk that must be retired/mitigated during the life cycle of the program (primarily during the development portion of the life cycle) using CRM processes to track progress towards the program's goal. RIDM helps to ensure that decisions between alternatives are made with an awareness of the risks associated with each, thereby helping to prevent late design changes, which can be key drivers of risk, cost overruns, schedule delays, and cancellation. Most project cost-saving opportunities occur in the definition, planning, and early design phases of a project.

The RIDM process described in this document attempts to respond to some of the primary issues that have derailed programs in the past: namely 1) the "mismatch" between stakeholder expectations and the "true" resources required to address the risks to achieve those expectations, 2) the miscomprehension of the risk that a decision-maker is accepting when making commitments to stakeholders, and 3) the miscommunication in considering the respective risks associated with competing alternatives.

This handbook is primarily written for systems engineers, risk managers, and risk analysts assigned to apply the requirements of NPR 8000.4A, but program managers of NASA programs and projects can get a sense of the value added by the process by reading the "RIDM Overview" section. It is designed to provide a concise description of RIDM and highlight key areas of the process. It can also be easily applied by unit engineers for application to units under their purview, although the application at such a low level should be based on the complexity of the engineering issue being addressed.

The RIDM methodology introduced by this handbook is part of a systems engineering process which emphasizes the proper use of risk analysis in its broadest sense to make risk-informed decisions that impact all mission execution domains, including safety, technical, cost, and schedule. In future versions of this handbook, the risk management principles discussed here will be updated in an evolutionary manner and expanded to address operations procedures procurement, strategic planning, and institutional risk management as experience is gained in the field. Technical appendices will be developed and added to provide tools and templates for implementation of the RIDM process. Examples will continue to be developed and will be disseminated as completed.

This handbook has been informed by many other guidance efforts underway at NASA, including the NASA Systems Engineering Handbook (NASA/SP-2007-6105 Rev. 1), the 2008 NASA Cost Estimating Handbook (NASA CEH-2008), and the NASA Standard for Models and Simulation (NASA-STD-7009) to name a few. How these documents relate and interact with the RIDM Handbook is discussed in subsequent chapters. With this in mind, this handbook could be seen as a complement to those efforts in order to help ensure programmatic success. In fact, the RIDM methodology has been formulated to complement, but not duplicate, the guidance in those documents. Taken together the overall guidance is meant to maximize program/project success by providing systematic and well-thought-out processes for conducting the discipline processes as well as integrating them into a formal risk analysis framework and communicating those results to a decision-maker so that he or she can make the best-informed decision possible.

Lastly, although formal decision analysis methods are now highly developed for unitary decision-makers, it is still a significant challenge to apply these methods in a practical way within a complex organizational hierarchy having its own highly developed program management policies and practices. This handbook is a step towards meeting that challenge for NASA but certainly not the final step in realizing the proper balance between formalism and practicality. Therefore, efforts will continue to ensure that the methods in this document are properly integrated and updated as necessary, to provide value to the program and project management processes at NASA.

Homayoon Dezfuli, Ph.D.
Project Manager, NASA Headquarters
April 2010

RISK-INFORMED DECISION MAKING
In the Context of NASA Risk Management

1.   INTRODUCTION

1.1   Purpose

The purpose of this handbook is to provide guidance for implementing the risk-informed decision making (RIDM) requirements of NASA Procedural Requirements (NPR) document NPR 8000.4A, Agency Risk Management Procedural Requirements [1], with a specific focus on programs and projects in the Formulation phase, and applying to each level of the NASA organizational hierarchy as requirements flow down. Appendix A provides a cross-reference between the RIDM-related requirements in NPR 8000.4A and the sections of this handbook for which guidance is provided.

This handbook supports RIDM application within the NASA systems engineering process, and is a complement to the guidance contained in NASA/SP-2007-6105, NASA Systems Engineering Handbook [2]. Figure 1 shows where the specific processes from the discipline-oriented NPR 7123.1, NASA Systems Engineering Process and Requirements [3], and NPR 8000.4 intersect with product-oriented NPRs, such as NPR 7120.5D, NASA Space Flight Program and Project Management Requirements [4]; NPR 7120.7, NASA Information Technology and Institutional Infrastructure Program and Project Management Requirements [5]; and NPR 7120.8, NASA Research and Technology Program and Project Management Requirements [6]. In much the same way that the NASA Systems Engineering Handbook is intended to provide guidance on the specific systems engineering processes established by NPR 7123.1, this handbook is intended to provide guidance on the specific RIDM processes established by NPR 8000.4A.

1.2   Scope and Depth

This handbook provides guidance for conducting risk-informed decision making in the context of NASA risk management (RM), with a focus on the types of direction-setting key decisions that are characteristic of the NASA program and project life cycles, and which produce derived requirements in accordance with existing systems engineering practices that flow down through the NASA organizational hierarchy. The guidance in this handbook is not meant to be prescriptive. Instead, it is meant to be general enough, and contain a sufficient diversity of examples, to enable the reader to adapt the methods as needed to the particular decision problems that he or she faces. The handbook highlights major issues to consider when making decisions in the presence of potentially significant uncertainty, so that the user is better able to recognize and avoid pitfalls that might otherwise be experienced.

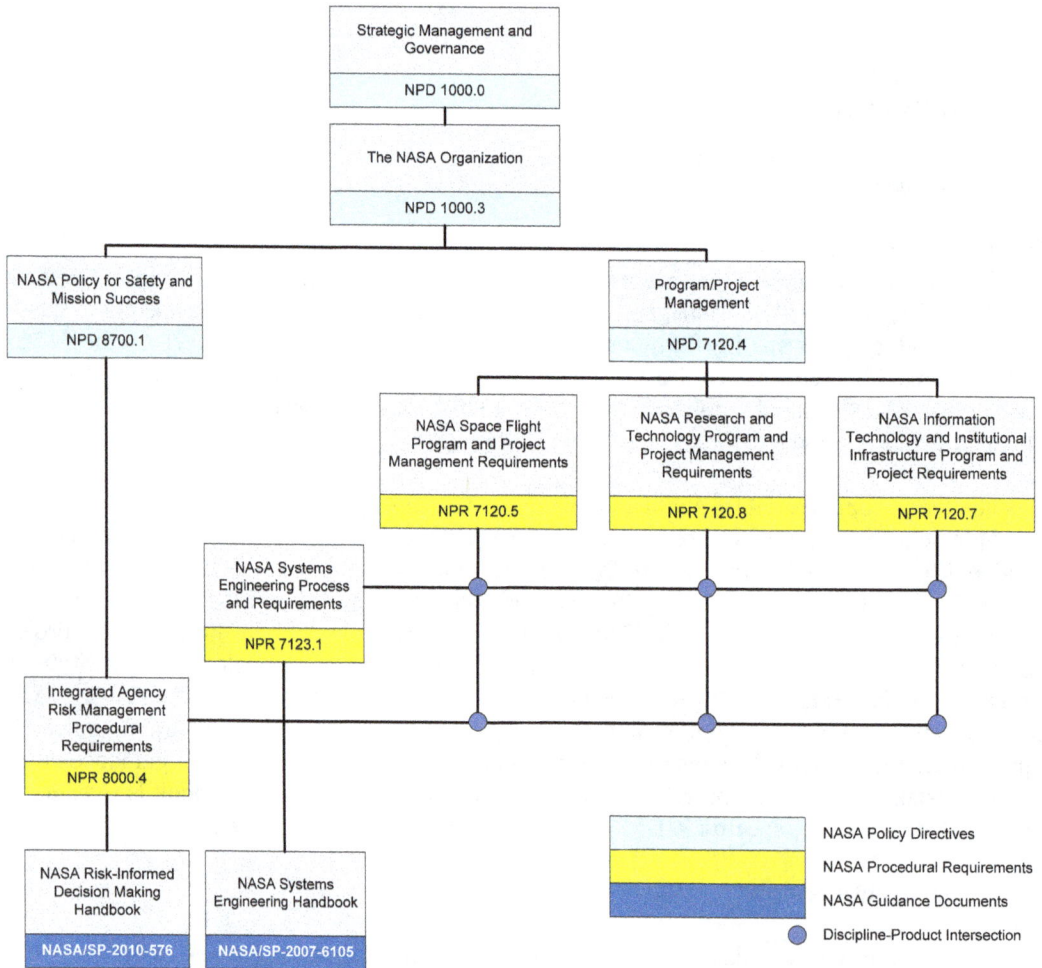

Figure 1. Intersection of Discipline Oriented and Product Oriented NPRs and their Associated Guidance Documents

Examples are provided throughout the handbook, and in Appendix F, to illustrate the application of RIDM methods to specific decisions of the type that are routinely encountered in NASA programs and projects. An example notional planetary mission is postulated and used throughout the document as a basis for illustrating the execution of the various process steps that constitute risk-informed decision making in a NASA risk management context ("yellow boxes"). In addition, key terms and concepts are defined throughout the document ("blue boxes").

Where applicable, guidance is also given on the spectrum of techniques that are appropriate to use, given the spectrum of circumstances under which decisions are made, ranging from narrow-scope decisions at the hardware component level that must be made using a minimum of time and resources, to broad-scope decisions involving multiple organizations upon which significant resources may be brought to bear. In all cases, the goal is to apply a level of effort to the task of risk-informed decision making that provides assurance that decisions are robust.

Additional guidance is planned to address more broadly the full scope of risk management requirements set forth in NPR 8000.4A, including:

- Implementation of the RIDM process in the context of institutional risk management; and

- Implementation of Continuous Risk Management (CRM) in conjunction with RIDM.

1.3   Background

NPR 8000.4A provides the requirements for risk management for the Agency, its institutions, and its programs and projects as required by NASA Policy Directive (NPD) 1000.5, Policy for NASA Acquisition [7]; NPD 7120.4C, Program/Project Management [8]; and NPD 8700.1, NASA Policy for Safety and Mission Success [9].

As discussed in NPR 8000.4A, risk is the potential for performance shortfalls, which may be realized in the future, with respect to achieving explicitly established and stated performance requirements. The performance shortfalls may be related to institutional support for mission execution[1] or related to any one or more of the following mission execution domains:

- Safety

- Technical

- Cost

- Schedule

Risk is operationally defined as a set of triplets:

- The *scenario(s)* leading to degraded performance with respect to one or more performance measures (e.g., scenarios leading to injury, fatality, destruction of key assets; scenarios leading to exceedance of mass limits; scenarios leading to cost overruns; scenarios leading to schedule slippage).

- The *likelihood(s)* (qualitative or quantitative) of those scenarios.

- The *consequence(s)* (qualitative or quantitative severity of the performance degradation) that would result if those scenarios were to occur.

---

[1] For the purposes of this version of the handbook, performance shortfalls related to institutional support for mission execution are subsumed under the affected mission execution domains of the program or project under consideration. More explicit consideration of institutional risks will be provided in future versions of this handbook.

Uncertainties are included in the evaluation of likelihoods and consequences.

Defining risk in this way supports risk management because:

- It distinguishes *high-probability*, *low-consequence* outcomes from *low-probability*, *high-consequence* outcomes;

- It points the way to proactive risk management controls, for example by supporting identification of risk drivers and the screening of *low-probability*, *low-consequence* outcomes; and

- It can point the way to areas where investment is warranted to reduce uncertainty.

In order to foster proactive risk management, NPR 8000.4A integrates two complementary processes, RIDM and CRM, into a single coherent framework. The RIDM process addresses the risk-informed selection of decision alternatives to assure effective approaches to achieving objectives, and the CRM process addresses implementation of the selected alternative to assure that requirements are met. These two aspects work together to assure effective risk management as NASA programs and projects are conceived, developed, and executed. Figure 2 illustrates the concept.

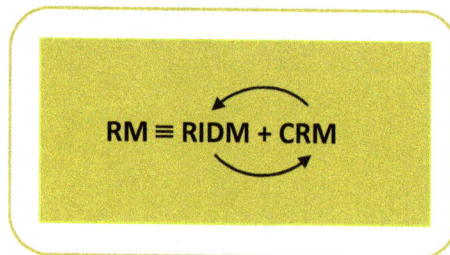

RM ≡ RIDM + CRM

Figure 2. Risk Management as the Interaction of Risk-Informed Decision Making and Continuous Risk Management

Risk-informed decision making is distinguished from risk-based decision making in that RIDM is a fundamentally deliberative process that uses a diverse set of performance measures, along with other considerations, to inform decision making. The RIDM process acknowledges the role that human judgment plays in decisions, and that technical information cannot be the sole basis for decision making. This is not only because of inevitable gaps in the technical information, but also because decision making is an inherently subjective, values-based enterprise. In the face of complex decision making involving multiple competing objectives, the cumulative wisdom provided by experienced personnel is essential for integrating technical and nontechnical factors to produce sound decisions.

Within the NASA organizational hierarchy, high-level objectives, in the form of NASA Strategic Goals, flow down in the form of progressively more detailed performance requirements, whose satisfaction assures that the objectives are met. Each organizational unit within NASA negotiates with the unit(s) at the next lower level in the organizational hierarchy a set of objectives, deliverables, performance measures, baseline performance requirements, resources, and schedules that defines the tasks to be performed by the unit(s). Once established, the lower level organizational unit manages its own risks against these specifications, and, as appropriate, reports risks and elevates decisions for managing risks to the next higher level based on predetermined risk thresholds that have been negotiated between the two units. Figure 3 depicts this concept. Invoking the RIDM process in support of key decisions as requirements flow down through the organizational hierarchy assures that objectives remain tied to NASA Strategic Goals while also capturing why a particular path for satisfying those requirements was chosen.

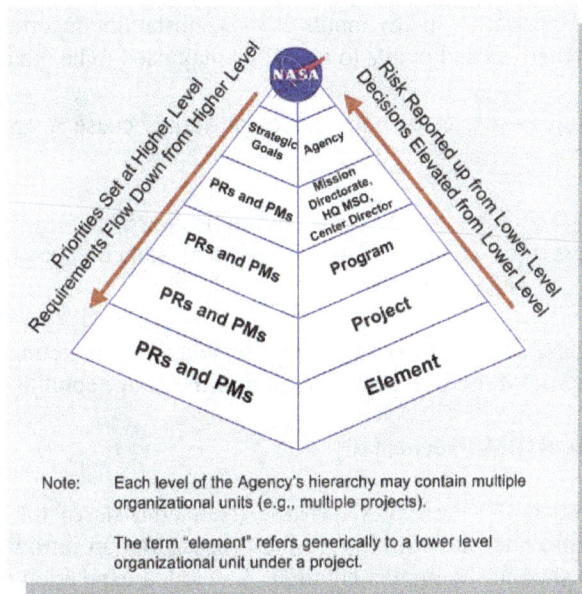

Figure 3. Flowdown of Performance Requirements (Illustrative)

1.4   When is RIDM Invoked?

RIDM is invoked for key decisions such as architecture and design decisions, make-buy decisions, source selection in major procurements, and budget reallocation (allocation of reserves), which typically involve requirements-setting or rebaseling of requirements.

RIDM is invoked in many different venues, based on the management processes of the implementing organizational unit. These include boards and panels, authority to proceed milestones, safety review boards, risk reviews, engineering design and operations planning decision forums, configuration management processes, and commit-to-flight reviews, among others.

RIDM is applicable throughout the project life cycle whenever trade studies are conducted. The processes for which decision analysis is typically appropriate, per Section 6.8.1 of the NASA Systems Engineering Handbook, are also those for which RIDM is typically appropriate. These decisions typically have one or more of the following characteristics:

- High Stakes — High stakes are involved in the decision, such as significant costs, significant potential safety impacts, or the importance of meeting the objectives.

- Complexity — The actual ramifications of alternatives are difficult to understand without detailed analysis.

- Uncertainty — Uncertainty in key inputs creates substantial uncertainty in the outcome of the decision alternatives and points to risks that may need to be managed.

- Multiple Attributes — Greater numbers of attributes cause a greater need for formal analysis.

- Diversity of Stakeholders — Extra attention is warranted to clarify objectives and formulate performance measures when the set of stakeholders reflects a diversity of values, preferences, and perspectives.

Satisfaction of all of these conditions is not a requirement for conducting RIDM. The point is, rather, that the need for RIDM increases as a function of the above conditions.

1.5   Overview of the RIDM Process [10]

As specified in NPR 8000.4A, the RIDM process itself consists of the three parts shown in Figure 4. This section provides an overview of the process and an introduction to the concepts and terminology established for its implementation. A detailed exposition of the steps associated with each part of the process can be found in Section 3, The RIDM Process.

Throughout the RIDM process, interactions take place between the *stakeholders*, the *risk analysts*, the *subject matter experts (SMEs)*, the *Technical Authorities*, and the *decision-maker* to ensure that objectives, values, and knowledge are properly integrated and communicated into the deliberations that inform the decision.

Figure 5 notionally illustrates the functional roles and internal interfaces of RIDM. As shown in the figure, it is imperative that the analysts conducting the risk analysis of alternatives incorporate the objectives of the various stakeholders into their analyses. These analyses are performed by, or with the support of, subject matter experts in the domains spanned by the objectives. The completed risk analyses are deliberated, along with other considerations, and the decision-maker selects a decision alternative for implementation (with the concurrence of the relevant Technical Authorities). The risk associated with the selected decision alternative becomes the central focus of CRM activities, which work to mitigate it during implementation, thus avoiding performance shortfalls in the outcome.

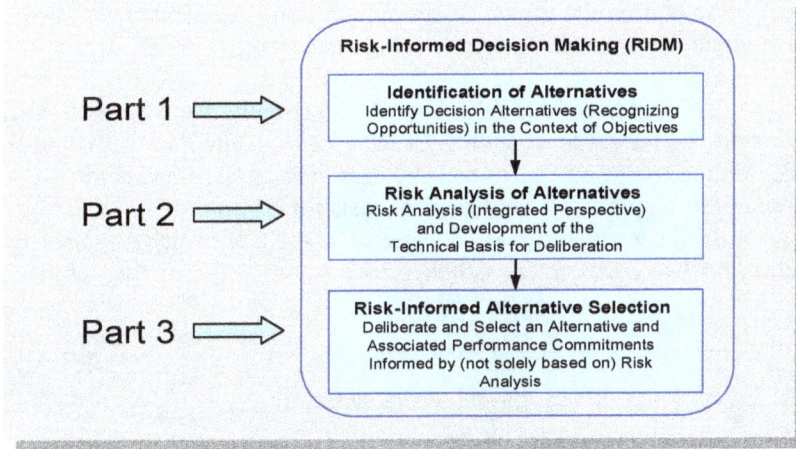

Figure 4. The RIDM Process

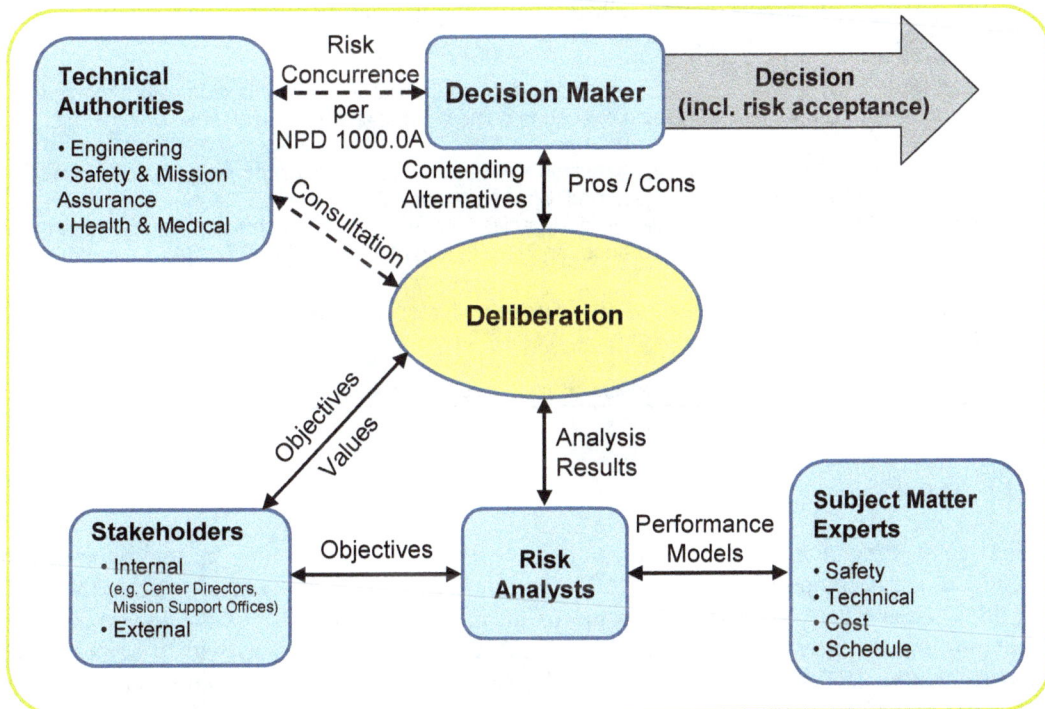

Figure 5. Functional Roles and Information Flow in RIDM (Notional)

The RIDM process is portrayed in this handbook primarily as a linear sequence of steps, each of which is conducted by individuals in their roles as stakeholders, risk analysts, subject matter experts, and decision-makers. The linear step-wise approach is used for instructional purposes only. In reality, some portions of the processes may be conducted in parallel, and steps may be iterated upon multiple times before moving to subsequent steps.

In particular, Part 2, Risk Analysis of Alternatives, is internally iterative as analyses are refined to meet decision needs in accordance with a graded approach, and Part 2 is iterative with Part 3, Risk-Informed Alternative Selection, as stakeholders and decision-makers iterate with the risk analysts in order to develop a sufficient technical basis for robust decision making. Additionally, decisions may be made via a series of downselects, each of which is made by a different decision-maker who has been given authority to act as proxy for the responsible decision authority.

RIDM Functional Roles*

Stakeholders - A stakeholder is an individual or organization that is materially affected by the outcome of a decision or deliverable but is outside the organization doing the work or making the decision [NPD 1000.0A]; e.g., Center Directors (CDs), Mission Support Offices (MSOs).

Risk Analysts – A risk analyst is an individual or organization that applies probabilistic methods to the quantification of performance with respect to the mission execution domains of safety, technical, cost, and schedule.

Subject Matter Experts – A subject matter expert is an individual or organization with expertise in one or more topics within the mission execution domains of safety, technical, cost, or schedule.

Technical Authorities – The individuals within the Technical Authority process who are funded independently of a program or project and who have formally delegated Technical Authority traceable to the Administrator. The three organizations who have Technical Authorities are Engineering, Safety and Mission Assurance, and Health and Medical. [NPD 1000.0A]

Decision-Maker – A decision-maker is an individual with responsibility for decision making within a particular organizational scope.

*Not to be interpreted as official job positions but as functional roles.

Part 1, Identification of Alternatives

In Part 1, *Identification of Alternatives,* objectives, which in general may be multifaceted and qualitative, are decomposed into their constituent-derived objectives, each of which reflects an individual issue that is significant to some or all of the stakeholders. At the lowest level of decomposition are **performance objectives**, each of which is associated with a **performance measure** that quantifies the degree to which the performance objective is addressed by a given decision alternative. In general, a performance measure has a "direction of goodness" that indicates the direction of increasingly beneficial performance measure values. A comprehensive set of performance measures is considered collectively for decision making, reflecting stakeholder interests and spanning the mission execution domains of:

- Safety (e.g., avoidance of injury, fatality, or destruction of key assets)

- Technical (e.g., thrust or output, amount of observational data acquired)

- Cost (e.g., execution within allocated cost)

- Schedule (e.g., meeting milestones)

Objectives whose performance measure values must remain within defined limits for every feasible decision alternative give rise to *imposed constraints* that reflect those limits. Objectives and imposed constraints form the basis around which decision alternatives are compiled, and performance measures are the means by which their ability to meet imposed constraints and satisfy objectives is quantified.

Part 2, Risk Analysis of Alternatives

In Part 2, *Risk Analysis of Alternatives*, the performance measures of each alternative are quantified, taking into account any significant uncertainties that stand between the selection of an the alternative and the accomplishment of the objectives. Given the presence of uncertainty, the actual outcome of a particular decision alternative will be only one of a spectrum of forecasted outcomes, depending on the occurrence, nonoccurrence, or quality of occurrence of intervening events. Therefore, it is incumbent on risk analysts to model each significant possible outcome, accounting for its probability of occurrence, in terms of the scenarios that produce it. This produces a distribution of outcomes for each alternative, as characterized by probability density functions (pdfs) over the performance measures (see Figure 6).

RIDM is conducted using a graded approach, i.e., the depth of analysis needs to be commensurate with the stakes and complexity of the decision situations being addressed. Risk analysts conduct RIDM at a level sufficient to support robust selection of a preferred decision alternative. If the uncertainty on one or more performance measures is preventing the decision-maker from confidently assessing important differences between alternatives, then the risk analysis may be iterated in an effort to reduce uncertainty. The analysis stops when the technical case is made; if the level of uncertainty does not preclude a *robust decision* from being made then no further uncertainty reduction is warranted.

Robustness

A robust decision is one that is based on sufficient technical evidence and characterization of uncertainties to determine that the selected alternative best reflects decision-maker preferences and values given the state of knowledge at the time of the decision, and is considered insensitive to credible modeling perturbations and realistically foreseeable new information.

The principal product of the risk analysis is the *Technical Basis for Deliberation (TBfD)*, a document that catalogues the set of candidate alternatives, summarizes the analysis methodologies used to quantify the performance measures, and presents the results. The TBfD is the input that risk-informs the deliberations that support decision making. The presence of this information does not necessarily mean that a decision is risk-informed; rather, without such information, a decision is not risk-informed. Appendix D contains a template that provides guidance on TBfD content. It is expected that the TBfD will evolve as the risk analysis iterates.

Performance Objectives, Performance Measures, and Imposed Constraints

In RIDM, top-level objectives, which may be multifaceted and qualitative, are decomposed into a set of **performance objectives**, each of which is implied by the top-level objectives, and which cumulatively encompass all the facets of the top-level objectives. Unlike top-level objectives, each performance objective relates to a single facet of the top-level objectives, and is quantifiable. These two properties of performance objectives enable quantitative comparison of decision alternatives in terms of capabilities that are meaningful to the RIDM participants. Examples of possible performance objectives are:

- Maintain Astronaut Health and Safety
- Minimize Cost

- Maximize Payload Capability
- Maximize Public Support

A performance measure is a metric used to quantify the extent to which a performance objective is fulfilled. In RIDM, a performance measure is associated with each performance objective, and it is through performance measure quantification that the capabilities of the proposed decision alternatives are assessed. Examples of possible performance measures, corresponding to the above performance objectives, are:

- Probability of Loss of Crew (P(LOC))
- Cost ($)

- Payload Capability (kg)
- Public Support (1 – 5)

Note that, in each case, the performance measure is the means by which the associated performance objective is assessed. For example, the ability of a proposed decision alternative to Maintain Astronaut Health and Safety (performance objective) may be measured in terms of its ability to minimize the Probability of Loss of Crew, P(LOC) (performance measure).

Although performance objectives relate to single facets of the top-level objectives, this does not necessarily mean that the corresponding performance measure is directly measurable. For example, P(LOC) might be used to quantify Maintain Astronaut Health and Safety, but the quantification itself might entail an assessment of vehicle reliability and abort effectiveness in the context of the defined mission profile.

An imposed constraint is a limit on the allowable values of the performance measure with which it is associated. Imposed constraints reflect performance requirements that are negotiated between NASA organizational units and which define the task to be performed. In order for a proposed decision alternative to be feasible it must comply with the imposed constraints. A hard limit on the minimum payload capability that is acceptable is an example of a possible imposed constraint.

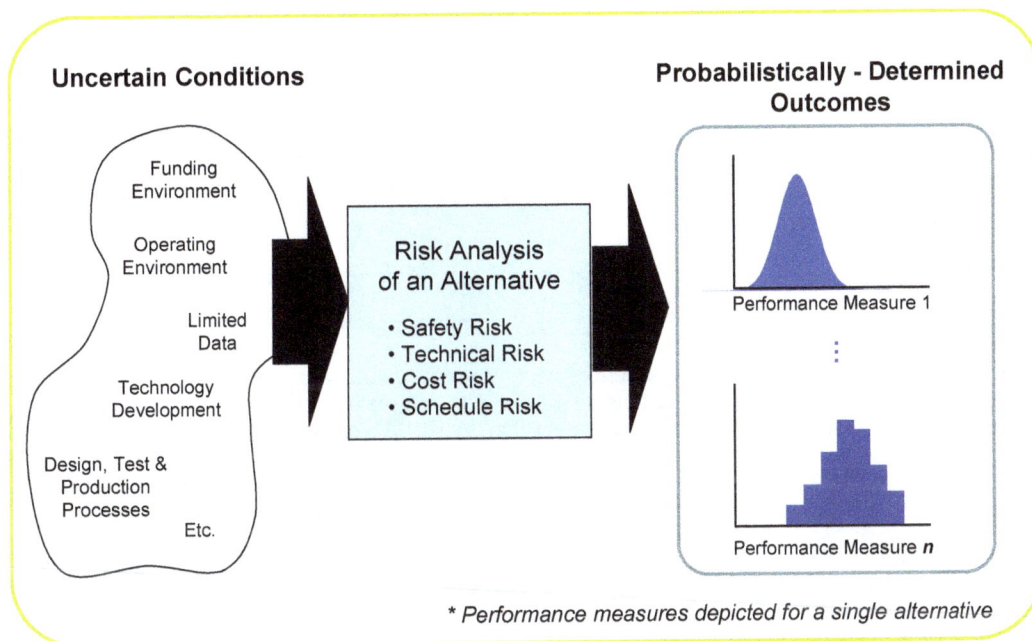

Figure 6. Uncertainty of Forecasted Outcomes Due to Uncertainty of Analyzed Conditions

Part 3, Risk-Informed Alternative Selection

In Part 3, *Risk-Informed Alternative Selection*, deliberation takes place among the stakeholders and the decision-maker, and the decision-maker either culls the set of alternatives and asks for further scrutiny of the remaining alternatives OR selects an alternative for implementation OR asks for new alternatives.

To facilitate deliberation, a set of ***performance commitments*** is associated with each alternative. Performance commitments identify the performance that an alternative is capable of, at a given probability of exceedance, or risk tolerance. By establishing a risk tolerance for each performance measure independent of the alternative, comparisons of performance among the alternatives can be made on a risk-normalized basis. In this way, stakeholders and decision-makers can deliberate the performance differences between alternatives at common levels of risk, instead of having to choose between complex combinations of performance and risk.

Deliberation and decision making might take place in a number of venues over a period of time or tiered in a sequence of downselects. The rationale for the selected decision alternative is documented in a Risk-Informed Selection Report (RISR), in light of:

- The risk deemed acceptable for each performance measure;

- The risk information contained in the TBfD; and

- The pros and cons of each contending decision alternative, as discussed during the deliberations.

Guidance for the RISR is provided in Appendix E. This assures that deliberations involve discussion of appropriate risk-related issues, and that they are adequately addressed and integrated into the decision rationale.

Performance Commitments

A **performance commitment** is the performance measure value, at a given risk tolerance level for that performance measure, acceptable to the decision-maker for the alternative that was selected. Performance commitments are used within the RIDM process in order to:

- Allow comparisons of decision alternatives in terms of performance capability at the specified risk tolerances of each performance measure (i.e., risk normalized).

- Serve as the starting point for requirements development, so that a linkage exists between the selected alternative, the risk tolerance of the decision-maker, and the requirements that define the objective to be accomplished. Performance commitments are not themselves performance requirements. Rather, performance commitments represent achievable levels of performance that are used to risk-inform the development of credible performance requirements as part of the overall systems engineering process.

The figure below shows a Performance Commitment C for Performance Measure X. Performance Measure X is characterized by a probability density function (pdf), due to uncertainties that affect the analyst's ability to forecast a precise value. The decision maker's risk tolerance level for not meeting Performance Commitment C is represented by the shaded area labeled "Risk".

When comparing alternatives, the decision maker looks for an alternative whose performance commitments meet the imposed constraints and which compares favorably to the other alternatives. Performance commitments are discussed in detail in Section 3.3.1.

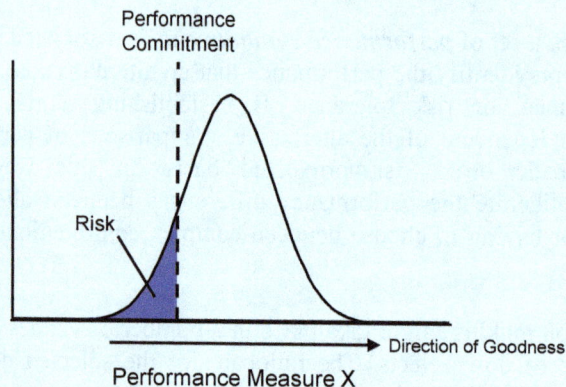

1.6 Avoiding Decision Traps

Examination of actual decision processes shows a tendency for decision-makers to fall into certain *decision traps*. These traps have been categorized as follows [11]:

- **Anchoring** — This trap is the tendency of decision-makers to give disproportionate weight to the first information they receive, or even the first hint that they receive. It is related to a tendency for people to reason in terms of perturbations from a "baseline" perception, and to formulate their baseline quickly and sometimes baselessly.

- **Status Quo Bias** — There is a tendency to want to preserve the status quo in weighing decision alternatives. In many decision situations, there are good reasons (e.g., financial) to preserve the status quo, but the bias cited here is a more basic tendency of the way in which people think. Reference [11] notes that early designs of "horseless carriages" were strongly based on horse-drawn buggies, despite being sub-optimal for engine-powered vehicles. There is also the tendency for managers to believe that if things go wrong with a decision, they are more likely to be punished for having taken positive action than for having allowed the status quo to continue to operate.

- **Sunk-Cost** — This refers to the tendency to throw good money after bad: to try to recoup losses by continuing a course of action, even when the rational decision would be to walk away, based on the current state of knowledge. This bias is seen to operate in the perpetuation of projects that are floundering by any objective standard, to the point where additional investment diverts resources that would be better spent elsewhere. A decision process should, in general, be based on the current situation: what gain is expected from the expenditure being contemplated.

- **Confirmation Bias** — This refers to the tendency to give greater weight to evidence that confirms our prior views, and even to seek out such evidence preferentially.

- **Framing** — This refers to a class of biases that relate to the human tendency to respond to how a question is framed, regardless of the objective content of the question. People tend to be risk-averse when offered the possibility of a sure gain, and risk-seeking when presented with a sure loss. However, it is sometimes possible to describe a given situation either way, which can lead to very different assessments and subsequent decisions.

- **Overconfidence** — This refers to the widespread tendency to underestimate the uncertainty that is inherent in the current state of knowledge. While most "experts" will acknowledge the presence of uncertainty in their assessments, they tend to do a poor job of estimating confidence intervals, in that the truth lies outside their assessed bounds much more often than would be implied by their stated confidence in those bounds. This is particularly true for decisions that are challenging to implement, as many decisions at NASA are. In the face of multiple sources of uncertainty, people tend to pay attention to the few with which they have the most experience, and neglect others. It is also particularly true for highly unlikely events, where there is limited data available to inform expert judgment.

- **Recallability** — This refers to the tendency of people to be strongly influenced by experiences or events that are easier for them to recall, even if a neutral statistical analysis of experience would yield a different answer. This means that dramatic or extreme events may play an unwarrantedly large role in decision making based on experience.

The RIDM process helps to avoid such traps by establishing a rational basis for decision-making, ensuring that the implications of each decision alternative have been adequately analyzed, and by providing a structured environment for deliberation in which each deliberator can express the merits and drawbacks of each alternative in light of the risk analysis results.

2.   RIDM PROCESS INTERFACES

As discussed in Section 1, within each NASA organizational unit, RIDM and CRM are integrated into a coherent RM framework in order to:

- Foster proactive risk management;

- Better inform decision making through better use of risk information; and

- More effectively manage implementation risks by focusing the CRM process on the baseline performance requirements emerging from the RIDM process.

The result is a RIDM process within each unit that interfaces with the unit(s) at the next higher and lower levels in the organizational hierarchy when negotiating objectives and establishing baseline performance requirements, as well as with its own unit's CRM process during implementation. This situation is illustrated graphically in Figure 7, which has been reproduced from NPR 8000.4A.[2] The following subsections discuss these interfaces in more detail.

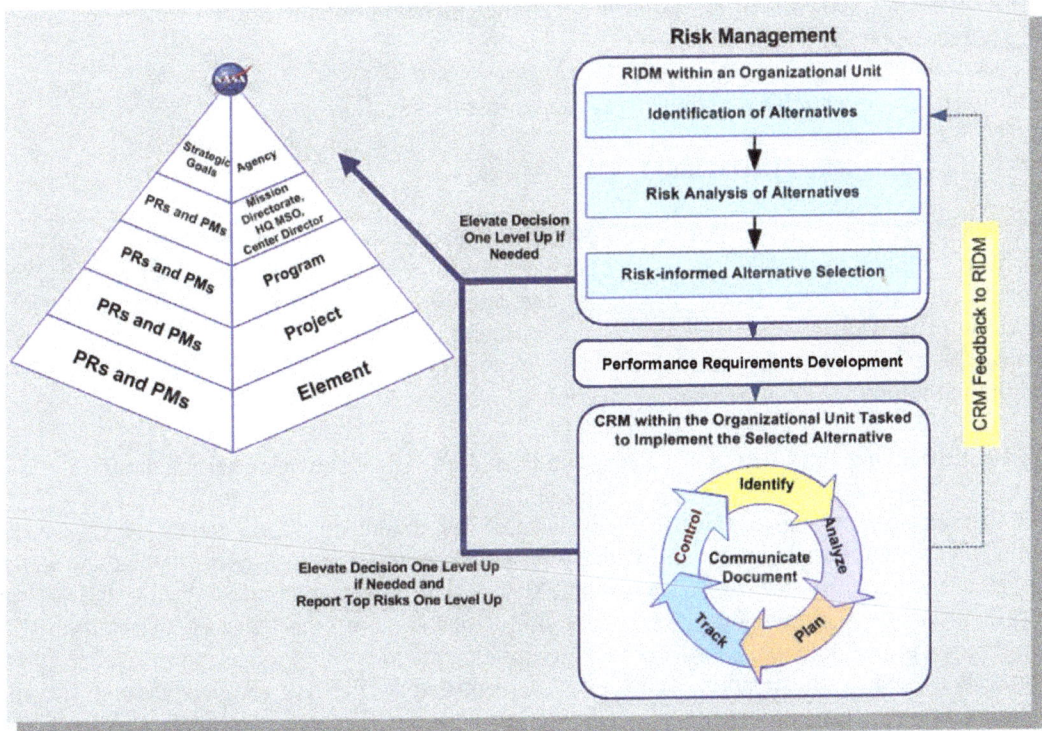

Figure 7. Coordination of RIDM and CRM Within the NASA Hierarchy (Illustrative)

---

[2] Figure 5 in NPR 8000.4A.

2.1   Negotiating Objectives Across Organizational Unit Boundaries

Organizational units negotiate with the unit(s) at the next lower level, including center support units, a set of objectives, deliverables, performance requirements, performance measures, resources, and schedules that defines the tasks to be performed. These elements reflect the outcome of the RIDM process that has been conducted by the level above and the execution of its own responsibility to meet the objectives to which it is working.

- The organizational unit at the level above is responsible for assuring that the objectives and imposed constraints assigned to the organizational unit at the lower level reflect appropriate tradeoffs between and among competing objectives and risks. Operationally, this means that a linkage is maintained to the performance objectives used in the RIDM process of the unit at the higher level. It also means that the rationale for the selected alternative is preserved, in terms of the imposed constraints that are accepted by the unit at the lower level.

- The organizational unit at the level below is responsible for establishing the feasibility and capability of accomplishing the objectives within the imposed constraints, and managing the risks of the job it is accepting (including identification of mission support requirements).

Additional discussion related to objectives can be found in Section 3.1.1.

2.2   Coordination of RIDM and CRM

RIDM and CRM are complementary RM processes that operate within every organizational unit. Each unit applies the RIDM process to decide how to meet objectives and applies the CRM process to manage risks associated with implementation.[3] In this way, RIDM and CRM work together to provide comprehensive risk management throughout the entire life cycle of the project. The following subsections provide an overview of the coordination of RIDM and CRM. Additional information can be found in Section 4.

2.2.1   Initializing the CRM Risks Using the Risk Analysis of the Selected Alternative

For the selected alternative, the risk analysis that was conducted during RIDM represents an initial identification and assessment of the scenarios that could lead to performance shortfalls. These scenarios form the basis for an initial risk list that is compiled during RIDM for consideration by the decision-maker. Upon implementation of the selected alternative, this information is available to the CRM process to initialize its *Identify, Analyze,* and *Plan* activities. Figure 8 illustrates the situation. The scenarios identified by the risk analysis are input to the *Identify* activity. The effects that these scenarios have on the ability to meet the baselined performance requirements are assessed in the *Analyze* activity. This activity integrates the

---

[3] In the context of CRM, the term "risk" is used to refer to a family of scenarios potentiated by a particular identifiable underlying condition that warrants risk management attention, because it can lead to performance shortfalls. This usage is more specific than the operational definition of risk presented in Section 1.3, and is formulated so that the underlying conditions can be addressed during implementation.

scenario-based risk analysis from RIDM into the CRM analysis activities as a whole, in the context of the baselined performance requirements to which CRM is managing. Strategies for addressing risks and removing threats to requirements are developed in the *Plan* activity, and are also informed by the RIDM risk analysis.

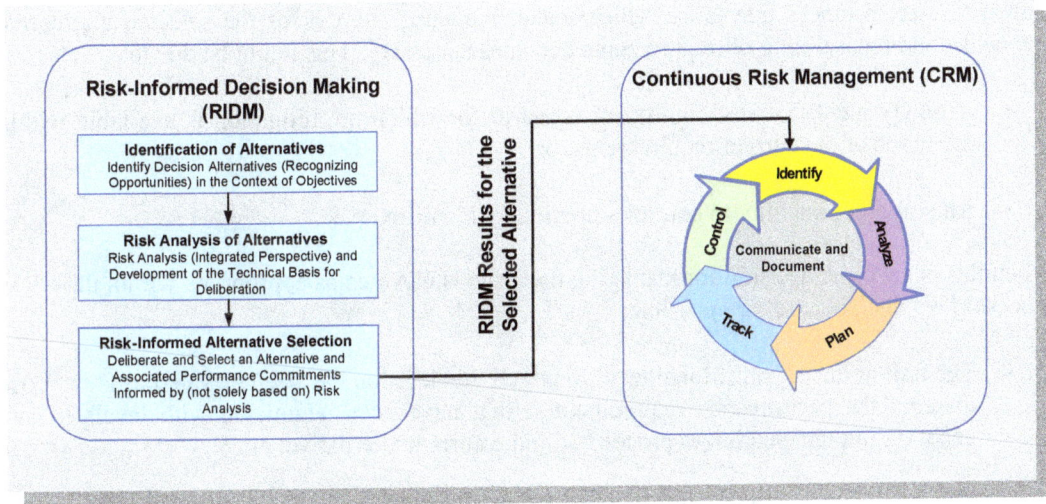

Figure 8. RIDM Input to CRM Initialization

While the risk analysis of the selected alternative *informs* CRM, it does not replace the need for independent CRM *Identify, Analyze,* and *Plan* activities. There are many reasons for this, but one key reason is that the risk analysis in RIDM is conducted expressly for the purposes of distinguishing between alternatives and generating performance commitments, not for the purpose of managing risk during implementation. Therefore, for example, uncertainties that are common to all alternatives and that do not significantly challenge imposed constraints will typically not be modeled to a high level of detail since they do not serve to discriminate between alternatives or affect the feasibility of the alternative. They will instead be modeled in a more simple and conservative manner. Also, the performance requirements of the selected alternative are baselined outside the RIDM process, and may differ from the performance commitments used in the risk analysis to define risk and develop mitigation strategies.

Once the CRM process produces a baseline risk list and develops mitigation strategies, these CRM products can be used to update the RIDM risk analysis for the selected alternative, as well as other alternatives to which the updated risk information and/or mitigation strategies are applicable. A change in the risk analysis results may represent an opportunity to reconsider the decision in light of the new information, and could justify modifying the selected alternative. Such opportunities can arise from any number of sources throughout the program/project life cycle. This feedback is illustrated in Figure 7.

2.2.2 Rebaselining of Performance Requirements

Following the selection of an alternative and the subsequent baselining of performance requirements, CRM operates to implement the selection in compliance with the performance requirements. Ideally, CRM will operate smoothly to achieve the objectives without incident. However, circumstances may arise which make managing the risk of the selected alternative untenable, and rebaselining of requirements becomes necessary. This might be due to:

- A newly identified risk-significant scenario for which no mitigation is available within the scope of the current requirements; or

- An emerging inability to control a previously identified risk.

When this occurs, the decision for managing the issue is elevated as appropriate within the CRM process. Two distinct cases are possible:

- The unit at the organizational level to which the decision has been elevated might choose to relax the performance requirement(s) that have been negotiated with levels below, enabling implementation to proceed with the current alternative, or

- The unit at the organizational level to which the decision has been elevated might choose not to modify the performance requirement(s) that have been negotiated with the unit that has elevated the decision. In this case, RIDM is re-executed at the level that must adhere to its performance requirements, potentially producing a new or modified alternative with corresponding new sets of derived performance requirements.

Rebaselining is done in light of current conditions. These conditions include not only the circumstances driving the rebaselining, but also those of the activity in general, such as budget status and accomplishments to date. The situation is shown in Figure 9, where decisions to address risks within CRM have been elevated to the appropriate level and RIDM is re-invoked to produce an updated alternative to serve as the basis for rebaselined requirements.

As indicated by the figure, the RIDM process is re-entered at Part 1, Identification of Alternatives, which addresses the development of performance measures and imposed constraints, as well as the compilation of a set of alternatives for analysis. In general, it is not expected that the performance measures will change, so RIDM is re-executed using those derived from the existing objectives hierarchy. However, there may be cause to modify an imposed constraint, particularly if it relates to the threatened requirement(s) and if modification/relaxation produces feasible requirements at a tolerable impact to objectives.

The set of decision alternatives compiled for risk analysis may also differ from the set analyzed initially. Alternatives that were previously shown to be unattractive can be excluded if they are unaffected by the circumstances surrounding the rebaselining. But the circumstances might also suggest alternatives that weren't considered before; care should be taken to identify these alternatives, and not draw only from the previous set.

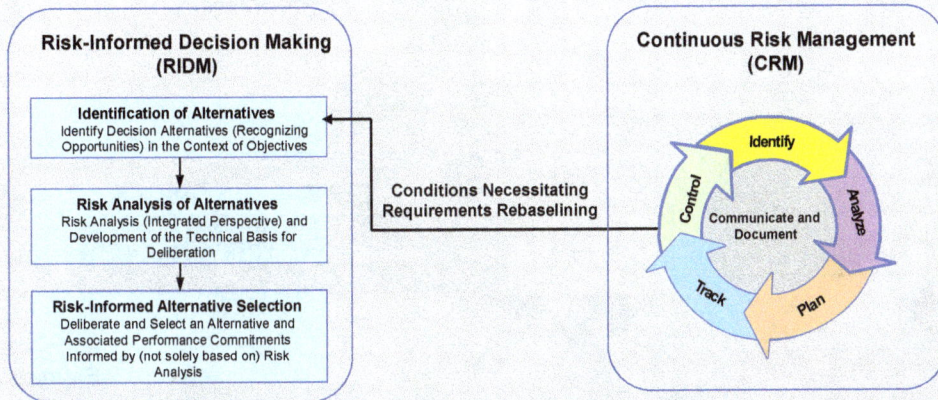

Figure 9. Rebaselining of Performance Requirements

Once the new set of decision alternatives is identified, RIDM proceeds as usual, taking advantage of the previous risk analysis to the extent practical given the new set and the current program/project status. Generation of a revised risk analysis to risk-inform subsequent deliberations will help the stakeholders and decision-maker to guard against sunk cost or *status quo* decision traps (discussed in Section 1.6) when deliberating and selecting a new or modified decision alternative.

The effects of requirements rebaselining are not confined to the organization that is the keeper of the original CRM risk. Every organization in the NASA hierarchy whose requirements are derived from the rebaselined requirements is potentially affected. The scope of affected organizations depends on the level at which the risk originates, the number of levels that the risk management decision is elevated by before it can be mitigated within the baseline requirements of the unit to which it is elevated, and the particulars of any changes to the mitigating unit's flowed-down requirements. Figure 10 illustrates the process, as well as the potential scope of the rebaselined requirements flowdown.

In certain instances, new information may emerge that represents an opportunity to rethink a previous decision. Just such a situation was mentioned in the case where the CRM process produces a mitigation strategy that, if retroactively applied to the set of candidate decision alternatives, could shift the preferred alternative from the selected alternative to a different one. Other opportunities can arise from ancillary analyses conducted either internally or externally to NASA, technology advancements, test results, etc.

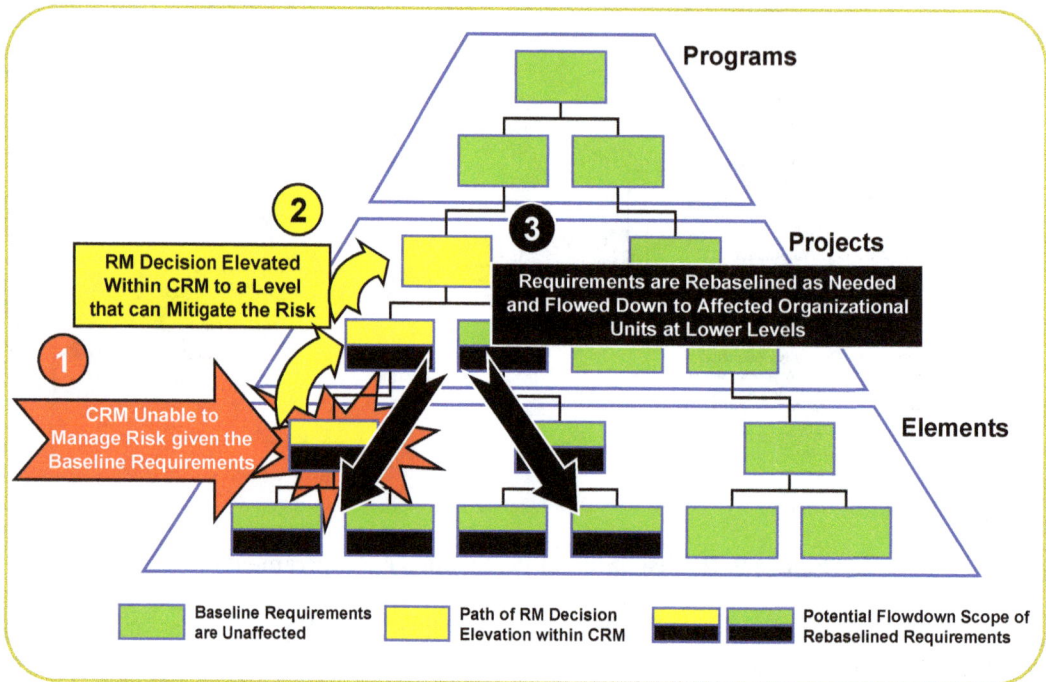

Figure 10. Scope of Potentially Affected Organizations Given Rebaselining

2.3   Maintaining the RIDM Process

The discussion of RIDM interfaces in the previous sections shows the importance of maintaining a functioning RIDM capability throughout the program/project life cycle. This capability includes:

- Reviewable TBfD and RISR documents containing the rationale for prior decision making and the discussion of issues of significance to the stakeholders.

- Accessible *objectives hierarchies* (discussed in Section 3.1.1) to serve as the sources of relevant performance measures or as the anchor points for decomposing objectives to finer levels of resolution. This assures that decisions remain tied to NASA strategic goals.

- Accessible risk analysis framework structures and risk models that were used to quantify the performance measures.

- The ability, at every organizational level, to integrate information from lower levels to support RIDM processes that reflect current conditions throughout the NASA hierarchy. This includes program/project status details as well as relevant analyses.

- Access to relevant discipline-specific analyses to use as input to risk analysis, as well as access to relevant expertise to support additional discipline-specific analyses needed for decision making.

- Maintenance of risk analysis expertise to coordinate the development of risk information and integrate it into the TBfD.

3.   THE RIDM PROCESS

Figure 11 expands the three parts of RIDM into a sequence of six process steps.

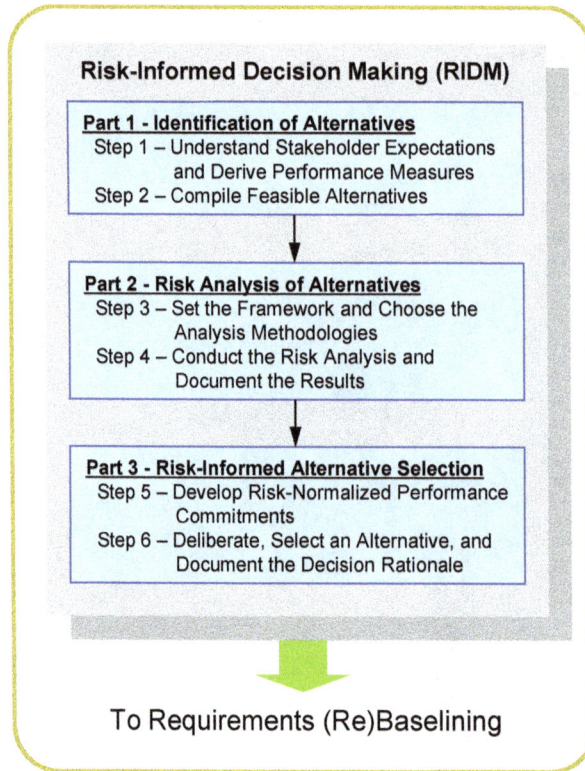

Risk-Informed Decision Making (RIDM)

Part 1 - Identification of Alternatives
Step 1 – Understand Stakeholder Expectations
           and Derive Performance Measures
Step 2 – Compile Feasible Alternatives

Part 2 - Risk Analysis of Alternatives
Step 3 – Set the Framework and Choose the
           Analysis Methodologies
Step 4 – Conduct the Risk Analysis and
           Document the Results

Part 3 - Risk-Informed Alternative Selection
Step 5 – Develop Risk-Normalized Performance
           Commitments
Step 6 – Deliberate, Select an Alternative, and
           Document the Decision Rationale

To Requirements (Re)Baselining

Figure 11. RIDM Process Steps

It is important to note that although Figures 4 and 11 depict the RIDM process as a linear sequence of steps, in practice it is expected that some steps could overlap in time and that the process is iterative. Information from latter steps feeds back into progressively more refined execution of previous steps until stakeholder issues are adequately addressed and the decision-maker has sufficient information, at a sufficient level of analytical rigor, to make a robust risk-informed decision. The primary issues driving the need for iteration are discussed in the following subsections, in the context of the RIDM process steps in which they arise.

The RIDM process has been informed by current theoretical and practical work in decision analysis and analytic-deliberative processes (see, for example, [12], [13], [14]). Some methodological tools and techniques, such as objectives hierarchies, performance measures, and deliberation, have been adopted into the RIDM process as being generally applicable to structured, rational decision making. Others, such as analytic hierarchy process (AHP) and multi-attribute utility theory (MAUT), are formally applicable to rational decision making but also present practical challenges in the context of requirements development within a complex organizational hierarchy having its own highly developed program management policies and practices. It is left to the discretion of the practitioner to determine, on a case-by-case basis, whether or not such techniques will aid in deliberation and selection of a decision alternative.

Planetary Science Mission Example

An example application of the RIDM process steps is presented in this handbook for a hypothetical planetary science mission. This example is contained in yellow boxes distributed throughout Section 3. Discussion of each RIDM process step is followed by a notional example of how it might be applied to a specific decision, derived from specific objectives.

The methods, measures, scope, and level of detail used in the planetary science mission example are not meant to prescribe how the RIDM process is to be applied in every instance. Rather, they are meant to give the reader a more concrete understanding of the RIDM process and its practical application, in addition to the more academic treatment in the main text.

The sections that follow provide a process overview, discussing each of the main activities that support each step.

3.1   Part 1 – Identification of Alternatives

As indicated in NPR 8000.4A and illustrated in Figure 4 of this handbook, decision alternatives are identifiable only in the context of the objectives they are meant to satisfy. Therefore, identification of alternatives begins with the process of understanding stakeholder expectations. From there, a basis for evaluating decision alternatives is developed by decomposing stakeholder expectations into quantifiable objectives that enable comparison among the candidates. Only then, after an appropriate context has been established, is it possible to compile a set of feasible alternatives that address the objectives. Figure 12 illustrates this part of the process, which is delineated in subsequent subsections.

3.1.1   Step 1 – Understand Stakeholder Expectations and Derive Performance Measures

3.1.1.1 Understand Stakeholder Expectations

The development of unambiguous performance measures and imposed constraints, reflecting stakeholder expectations, is the foundation of sound decision making. Paragraph 3.2.1 of NPR 7123.1A establishes systems engineering process requirements for stakeholder expectations definition, and Section 4.1 of the NASA Systems Engineering Handbook provides further guidance on understanding stakeholder expectations.

Part 1 – Identification of Alternatives

Begin RIDM Process

Output of Part 1
- Feasible alternatives
- Performance measures
- Imposed constraints

Step 1 - Understand Stakeholder Expectations and Derive Performance Measures

- Identify stakeholders and get input
- Negotiate flowed-down requirements
- Define top-level objectives
- Develop top-level boundaries and milestones
- Construct objectives hierarchy
- Derive performance objectives from top-level objectives
- Develop performance measures and imposed constraints

Step 2 - Compile Feasible Alternatives

- Get stakeholder input
- Construct trade tree of candidate alternatives
- Perform preliminary evaluation
- Prune infeasible alternatives

No

Do the performance measures and imposed constraints capture the objectives?

Yes

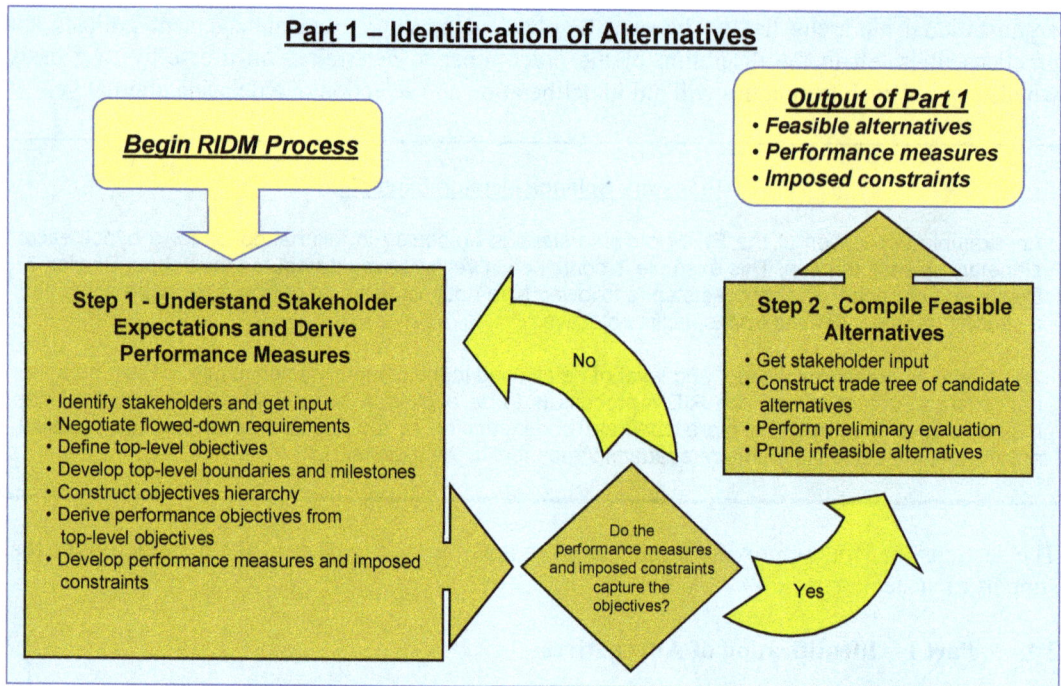

Figure 12. RIDM Process Flowchart: Part 1, Identification of Alternatives

Typical inputs needed for the stakeholder expectations definition process include:

- **Upper Level Requirements and Expectations:** These would be the requirements and expectations (e.g., needs, wants, desires, capabilities, constraints, external interfaces) that are being flowed down to a particular system of interest from a higher level (e.g., program, project, etc.).

- **Stakeholders:** Individuals or organizations that are materially affected by the outcome of a decision or deliverable but are outside the organization doing the work or making the decision.

A variety of organizations, both internal and external to NASA, may have a stake in a particular decision. Internal stakeholders might include NASA Headquarters (HQ), the NASA Centers, and NASA advisory committees. External stakeholders might include the White House, Congress, the National Academy of Sciences, the National Space Council, and many other groups in the science and space communities.

Stakeholder expectations, the vision of a particular stakeholder individual or group, result when they specify what is desired as an end state or as an item to be produced and put bounds upon the achievement of the goals. These bounds may encompass expenditures (resources), time to deliver, performance objectives, or other less obvious quantities such as organizational needs or geopolitical goals.

Typical outputs for capturing stakeholder expectations include the following:

- **Top-Level Requirements and Expectations:** These would be the top-level needs, wants, desires, capabilities, constraints, and external interfaces for the product(s) to be developed.

- **Top-Level Conceptual Boundaries and Functional Milestones:** This describes how the system will be operated during the life cycle phases to meet stakeholder expectations. It describes the system characteristics from an operational perspective and helps facilitate an understanding of the system goals. This is usually accomplished through use-case scenarios, design reference missions (DRMs), and concepts of operation (ConOps).

In the terminology of RIDM, the stakeholder expectations that are the outputs of this step consist of top-level objectives and imposed constraints. Top-level objectives state what the stakeholders hope to achieve from the activity. They are typically qualitative and multifaceted, reflecting competing sub-objectives (e.g., more data vs. lower cost). Imposed constraints represent the top-level success criteria for the undertaking, outside of which the top-level objectives are not achieved. For example, if an objective is to put a satellite of a certain mass into a certain orbit, then the ability to loft that mass into that orbit is an imposed constraint, and any proposed solution that is incapable of doing so is infeasible.

3.1.1.2 Derive Performance Measures

In general, decision alternatives cannot be directly assessed relative to multifaceted and/or qualitative top-level objectives. Although the top-level objectives state the goal to be accomplished, they may be too complex, as well as vague, for any operational purpose. To deal with this situation, objectives are decomposed, using an *objectives hierarchy*, into a set of conceptually distinct lower-level objectives that describe the full spectrum of necessary and/or desirable characteristics that any feasible and attractive alternative should have. When these objectives are quantifiable via performance measures, they provide a basis for comparing proposed alternatives.

Constructing an Objectives Hierarchy

An objectives hierarchy is constructed by subdividing an objective into lower-level objectives of more detail, thus clarifying the intended meaning of the general objective. Decomposing an objective into precise lower-level objectives clarifies the tasks that must be collectively achieved and provides a well-defined basis for distinguishing between alternative means of achieving them.

Planetary Science Mission Example: Understand Stakeholder Expectations

The Planet "X" Program Office established an objective of placing a scientific platform in orbit around Planet "X" in order to gather data and transmit it back to Earth. Stakeholders include:

- The planetary science community who will use the data to further humanity's understanding of the formation of the solar system

- The Earth science community who will use the data to refine models of terrestrial climate change and geological evolution

- Environmental groups who are concerned about possible radiological contamination of Planet "X" in the event of an orbital insertion mishap

- Mission support offices who are interested in maintaining their infrastructure and workforce capabilities in their areas of specialized expertise

Specific expectations include:

- The envisioned concept of operations is for a single launch of a scientific platform that will be placed in a polar orbit around Planet X

- The envisioned scientific platform will include a radioisotope thermoelectric generator (RTG) for electrical power generation

- The launch date must be within the next 55 months due to the launch window

- The scientific platform should provide at least 6 months of data collection

- Data collection beyond the initial 6 months is desirable but not mission critical

- The scientific platform will include a core data collection capability in terms of data type and data quality (for the purpose of this example, the specifics of the data are unspecified)

- Collection of additional (unspecified) types of scientific data is desirable if the capability can be provided without undue additional costs or mission success impacts

- The mission should be as inexpensive as possible, with a cost cap of $500M

- The probability of radiological contamination of Planet "X" should be minimized, with a goal of no greater than 1 in 1000 (0.1%)

An objectives hierarchy is shown notionally in Figure 13. At the first level of decomposition the top-level objective is partitioned into the NPR 8000.4A mission execution domains of Safety, Technical, Cost, and Schedule. This enables each performance measure and, ultimately, performance requirement, to be identified as relating to a single domain. Below each of these domains the objectives are further decomposed into sub-objectives, which themselves are iteratively decomposed until appropriate quantifiable performance objectives are generated.

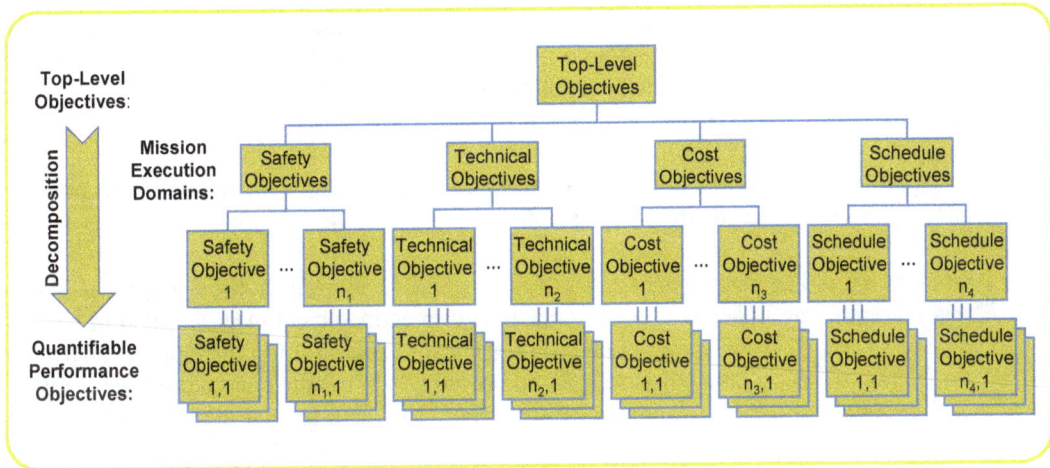

Figure 13. Notional Objectives Hierarchy

There is no prescribed depth to an objectives hierarchy, nor must all performance objectives reside at the same depth in the tree. The characteristics of an objectives hierarchy depend on the top-level objective and the context in which it is to be pursued. Furthermore, a unique objectives hierarchy is not implied by the specification of an objective; many different equally legitimate objectives hierarchies could be developed.

When developing an objectives hierarchy there is no obvious stopping point for the decomposition of objectives. Judgment must be used to decide where to stop by considering the advantages and disadvantages of further decomposition. Things to consider include:

- Are all facets of each objective accounted for?

- Are all the performance objectives at the levels of the hierarchy quantifiable?

- Is the number of performance objectives manageable within the scope of the decision-making activity?

One possibility is to use a "test of importance" to deal with the issue of how broadly and deeply to develop an objectives hierarchy and when to stop. Before an objective is included in the hierarchy, the decision-maker is asked whether he or she feels the best course of action could be altered if that objective were excluded. An affirmative response would obviously imply that the objective should be included. A negative response would be taken as sufficient reason for exclusion. It is important when using this method to avoid excluding a large set of attributes, each of which fails the test of importance but which collectively are important. As the decision-making process proceeds and further insight is gained, the test of importance can be repeated with the excluded objectives to assure that they remain non-determinative. Otherwise they must be added to the hierarchy and evaluated for further decomposition themselves until new stopping points are reached.

The decomposition of objectives stops when the set of performance objectives is operationally useful and quantifiable, and the decision-maker, in consultation with appropriate stakeholders, is satisfied that it captures the expectations contained in the top-level objective. It is desirable that the performance objectives have the following properties. They should be:

- Complete – The set of performance objectives is complete if it includes all areas of concern embedded in the top-level objective.

- Operational – The performance objectives must be meaningful to the decision-maker so that he or she can understand the implications of meeting or not meeting them to various degrees. The decision-maker must ultimately be able to articulate a rationale for preferring one decision alternative over all others, which requires that he or she be able to ascribe value, at least qualitatively, to the degree to which the various alternatives meet the performance objectives.

- Non-redundant – The set of performance objectives is non-redundant if no objective contains, or significantly overlaps with, another objective. This is not to say that the ability of a particular alternative to meet different performance objectives will not be correlated. For example, in application, *maximize reliability* is often negatively correlated with *minimize cost*. Rather, performance objectives should be conceptually distinct, regardless of any solution-specific performance dependencies.

- Solution independent – The set of performance objectives should be applicable to any reasonable decision alternative and should not presuppose any particular aspect of an alternative to the exclusion of other reasonable alternatives. For example, an objectives hierarchy for a payload launch capability that had *Minimize Slag Formation* as a performance objective would be presupposing a solid propellant design. Unless solid propellant was specifically required based on a prior higher-level decision, *Minimize Slag Formation* would not reflect an unbiased decomposition of the top-level objective.

Guidance on developing objectives hierarchies can be found in Clemen [12] and Keeney and Raiffa [13], as well as on websites such as Comparative Risk Assessment Framework and Tools (CRAFT) [15].

Fundamental vs. Means Objectives

When developing an objectives hierarchy it is important to use ***fundamental objectives*** as opposed to ***means objectives***. Fundamental objectives represent *what* one wishes to accomplish, as opposed to means objectives, which represent *how* one might accomplish it. Objectives hierarchies decompose high-level fundamental objectives into their constituent parts (partitioning), such that the fundamental objectives at the lower level are those that are implied by the fundamental objective at the higher level. In contrast, means objectives indicate a particular way of accomplishing a higher-level objective. Assessment of decision alternatives in terms of fundamental objectives as opposed to means objectives represents a performance-based approach to decision making, as recommended by the Aerospace Safety Advisory Panel (ASAP) as emphasizing "early risk identification to guide design, thus enabling creative design approaches that might be more efficient, safer, or both." [16].

The difference between fundamental objectives and means objectives is illustrated in Figure 14, which shows an objectives hierarchy on the top and a means objectives network on the bottom. The first thing to notice is that the objectives hierarchy is just that, a hierarchy. Each level decomposes the previous level into a more detailed statement of what the objectives entail. The objective, *Maximize Safety*, is decomposed (by partitioning) into *Minimize Loss of Life*, *Minimize Serious Injuries*, and *Minimize Minor Injuries*. The three performance objectives explain what is meant by *Maximize Safety*, without presupposing a particular way of doing so.[4]

In contrast, the means objectives network is not a decomposition of objectives, which is why it is structured as a network instead of a hierarchy. The objective, *Educate Public about Safety*, does not explain what is meant by any one the higher-level objectives; instead, it is a way of accomplishing them. Other ways may be equally effective or even more so. Deterministic standards in general are means objectives, as they typically prescribe techniques and practices by which fundamental objectives, such as safety, will be achieved. Means objectives networks arise in another context in the RIDM process and are discussed further in Section 3.2.1.

Performance Measures

Once an objectives hierarchy is completed that decomposes the top-level objective into a complete set of quantifiable performance objectives, a performance measure is assigned to each as the metric by which its degree of fulfillment is quantified. In many, if not most cases the appropriate performance measure to use is self-evident from the objective. In other cases the choice may not be as clear, and work must be done in order to assure that the objective is not only quantifiable, but that the performance measure used to quantify it is adequately representative of the objective to begin with.

Objectives that have natural unit scales (e.g., *Minimize Cost*, *Maximize Payload*) are generally easy to associate with an appropriate performance measures (e.g., *Total Cost* or *Cost Overrun* [$], *Payload Mass* [kg]). Other objectives might not have an obvious or practical natural unit scale, thereby requiring the development of either a *constructed scale* or a *proxy performance measure*.

A constructed scale is typically appropriate for measuring objectives that are essentially subjective in character, or for which subjective or linguistic assessment is most appropriate. An

---

[4] NASA is currently developing quantitative *safety goals* and associated *thresholds* (akin to imposed constraints) that will be used to guide risk acceptance decisions. [17] An example of a quantitative safety goal would be: the risk to an astronaut from the ascent phase of a launch to LEO should be less than <a specified value>.

example of such an objective might be *Maximize Stakeholder Support*. Here, stakeholder support is the attribute being measured, but there is no natural measurement scale by which an objective assessment of stakeholder support can be made. Instead, it might be reasonable to construct a scale that supports subjective/linguistic assessment of stakeholder support (see Table 1). Constructed scales are also useful as a means of quantifying what is essentially qualitative information, thereby allowing it to be integrated into a quantitative risk analysis framework.

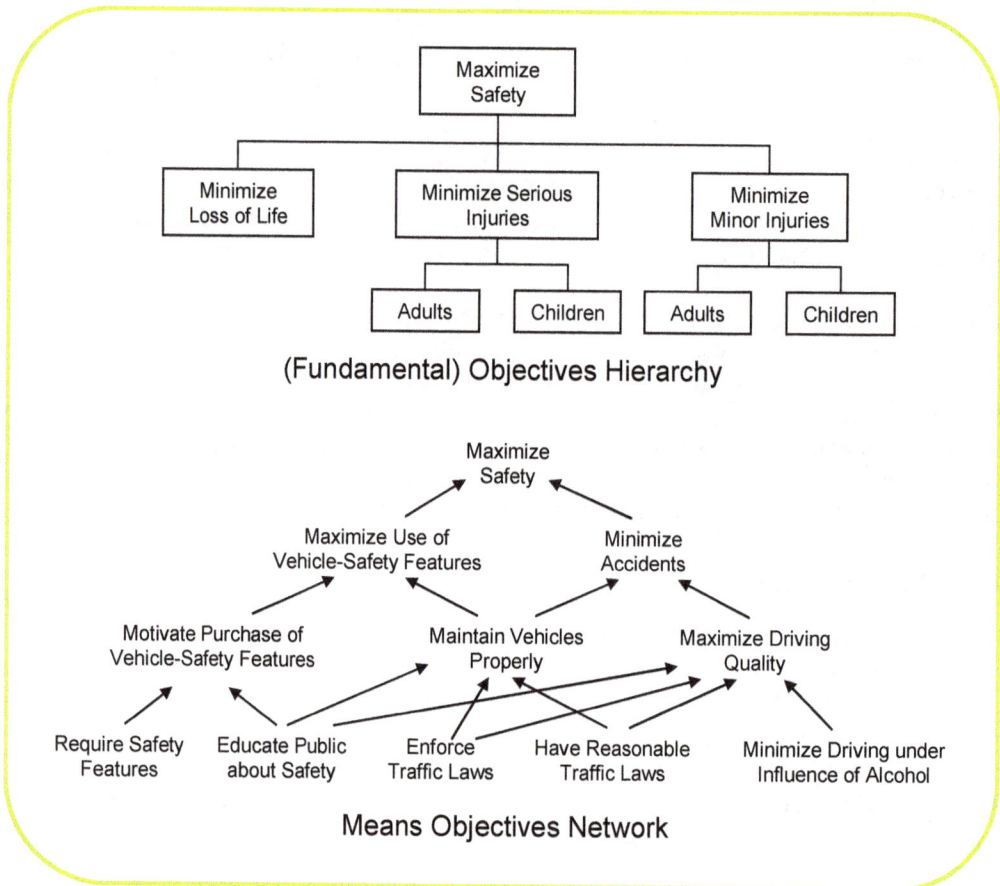

Figure 14. Fundamental vs. Means Objectives [19]

Alternatively, it may be possible to identify an objective performance measure that *indirectly* measures the degree of fulfillment of an objective. In the previous paragraph the objective, *Maximize Stakeholder Support*, was assessed subjectively using a *Stakeholder Support* performance measure with a constructed scale. Another strategy for assessing the objective might be to define a proxy for stakeholder support, such as the average number of stakeholders attending the bi-weekly status meetings. In this case, the proxy performance measure gives an indication of stakeholder support that might be operationally adequate for the decision at hand, although it does not necessarily correlate exactly to actual stakeholder support.

Table 1. A Constructed Scale for Stakeholder Support (Adapted from [12])

Scale	Value	Description
5	Action-oriented Support	Two or more stakeholders are actively advocating and no stakeholders are opposed.
4	Support	No stakeholders are opposed and at least one stakeholder has expressed support.
3	Neutrality	All stakeholders are indifferent or uninterested.
2	Opposition	One or more stakeholders have expressed opposition, although no stakeholder is actively opposing.
1	Action-oriented Opposition	One or more stakeholders are actively opposing.

The relationship between natural, constructed and proxy scales is illustrated in Figure 15 in terms of whether or not the performance measure directly or indirectly represents the corresponding objective, and whether the assessment is empirically quantifiable or must be subjectively assessed. Additionally, Figure 15 highlights the following two characteristics of performance measures:

- The choice of performance measure type (natural, constructed, proxy) is not a function of the performance measure alone. It is also a function of the performance objective that the performance measure is intended to quantify. For example P(LOC) can be considered a natural performance measure as applied to astronaut life safety, since it directly addresses astronaut casualty expectation. However, in some situations P(LOC) might be a good *proxy* performance measure for overall astronaut health, particularly in situations where astronaut injury and/or illness are not directly assessable.

- There is seldom, if ever, a need for an indirect, subjective performance measure. This is because performance objectives tend to be intrinsically amenable to direct, subjective assessment. Thus, for objectives that do not have natural measurement scales, it is generally productive to ask whether the objective is better assessed directly but subjectively, or whether it is better to forego direct measurement in exchange for an empirically-quantifiable proxy performance measure. The first case leads to a constructed performance measure that is direct but perhaps not reproducible; the second to a performance measure that is reproducible but may not fully address the corresponding performance objective.

A performance measure should be adequate in indicating the degree to which the associated performance objective is met. This is generally not a problem for performance measures that have natural or constructed scales, but can be a challenge for proxy performance measures. In the *Maximize Stakeholder Support* example above, it is possible that a stakeholder who perceives the activity to be an obstacle to his or her real objectives might attend the meetings in order to remain informed about potential threats. Thus the average number of stakeholders attending the status meetings might not be an accurate representation of stakeholder support, and in this case may have a contraindicative element to it.

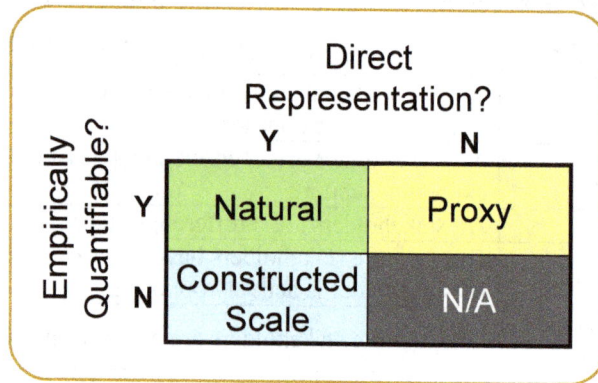

Figure 15. Types of Performance Measures

Figure 16 illustrates the relationship between performance objectives and performance measures. A performance measure has been established on each of the performance objectives based on the objective's natural measurement scale, a constructed scale that has been developed for subjective quantification, or via a proxy performance measure.

Although it is preferable that a performance measure be directly measurable, this is not always possible, even for objectives with natural measurement scales. For example, safety-related risk metrics such as *Probability of Loss of Mission, P(LOM)*, and *Probability of Loss of Crew, P(LOC)*, are typically used to quantify the objectives *Avoid Loss of Mission* and *Maintain Astronaut Health and Safety*. These performance measures are the product of modeling activities as opposed to direct measurement, involving the integration of numerous parameters within an analytical model of the alternative under consideration. In cases such as these, where modeling methods are integral to the resultant performance measure values, the modeling protocols become part of the performance measure definition. This assures that performance measures are calculated consistently.

One proxy performance measure of particular importance to many NASA decisions is *Flexibility*. *Flexibility* refers to the ability to support more than one current application. A technology choice that imposes a hard limit on the mass that can be boosted into a particular orbit has less flexibility than a choice that is more easily adaptable to boost more. The objective, *Maximize Flexibility*, allows this type of issue to be addressed systematically in decision making. However, since *Maximize Flexibility* refers to potential capabilities that are as yet undefined, there is no natural measurement scale that can be used for quantification.[5] A constructed scale is possible, although it requires subjective assessment. A proxy performance measure for flexibility can be constructed by, for example, assessing the capability of the alternative to support a selected set of alternative objectives, such as boosting a larger mass into orbit.

---

[5] In such applications, *Flexibility* is a surrogate for certain future performance attributes. This idea is discussed more extensively by Keeney [18] and Keeney and McDaniels [19].

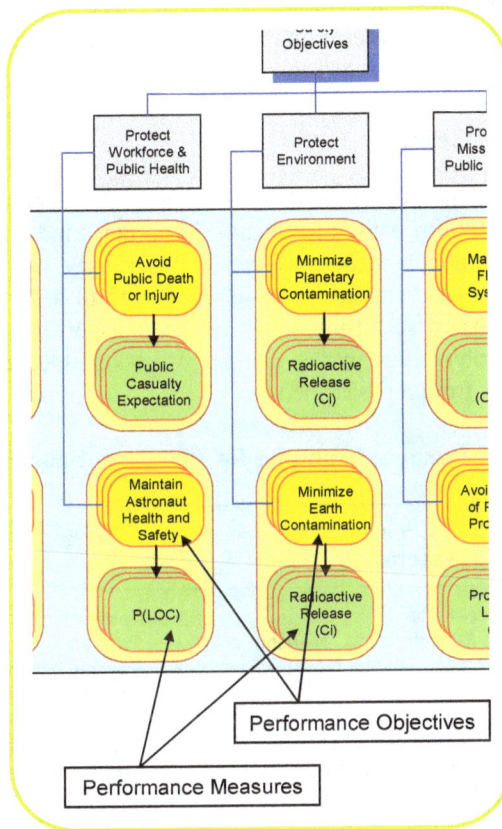

Figure 16. The Relationship between Performance Objectives and Performance Measures

Risk Minimization Is Not a Performance Objective

It is sometimes the practice in decision analyses and trade studies to treat *Minimize Risk* as a distinct performance objective, which is then decomposed into domains such as technology, programmatic, cost, and schedule, resulting in performance measures such as *technology risk, programmatic risk, cost risk,* and *schedule risk.* However, in NPR 8000.4A, risk is the potential for shortfalls with respect to performance requirements (which in a RIDM context translates operationally into shortfalls with respect to performance commitments). Therefore, *Minimize Risk* is not a distinct objective in the objectives hierarchy. Rather, it is the task of risk management itself (including RIDM), for which risk is an attribute of every performance objective, as measured by the probability of falling short of its associated performance commitment.

For example, if a certain payload capability is contingent on the successful development of a particular propulsion technology, then the risk of not meeting the payload performance commitment is determined in part by the probability that the technology development program will be unsuccessful. In other words, the risk associated with technology development is accounted for in terms of its risk impact on the performance commitments (in this case, payload). There is no need to evaluate a separate Technology Risk metric.[6]

Example Performance Measures

Performance measures should fall within the mission execution domains of safety, technical, cost and schedule. Table 2 contains a list of typically important kinds of performance measures for planetary spacecraft and launch vehicles. Note that this is by no means a comprehensive and complete list. Although such lists can serve as checklists to assure comprehensiveness of the derived performance measure set, it must be stressed that performance measures are explicitly derived from top-level objectives in the context of stakeholder expectations, and cannot be established prescriptively from a predefined set.

Table 2. Performance Measures Examples for Planetary Spacecraft and Launch Vehicles

Performance Measures for Planetary Spacecraft	Performance Measures for Launch Vehicles
• End-of-mission (EOM) dry mass • Injected mass (includes EOM dry mass, baseline consumables and upper stage adaptor mass) • Consumables at EOM • Power demand (relative to supply) • Onboard data processing memory demand • Onboard data processing throughput time • Onboard data bus capacity • Total pointing error	• Total vehicle mass at launch • Payload mass (at nominal altitude or orbit) • Payload volume • Injection accuracy • Launch reliability • In-flight reliability • For reusable vehicles, percent of value recovered • For expendable vehicles, unit production cost at the nth unit

---

[6] Unless *Engage in Technology Development* is a performance objective in its own right.

Planetary Science Mission Example: Derive Performance Measures

From the generic top-level objective of "Project Success", the stakeholder expectations that have been captured are organized via an objectives hierarchy that decomposes the top-level objective through the mission execution domains of Safety, Technical, Cost, and Schedule, producing a set of performance objectives at the leaves. The terminology of "Minimize" and "Maximize" is used as appropriate to indicate the "direction of goodness" that corresponds to increasing performance for that objective.

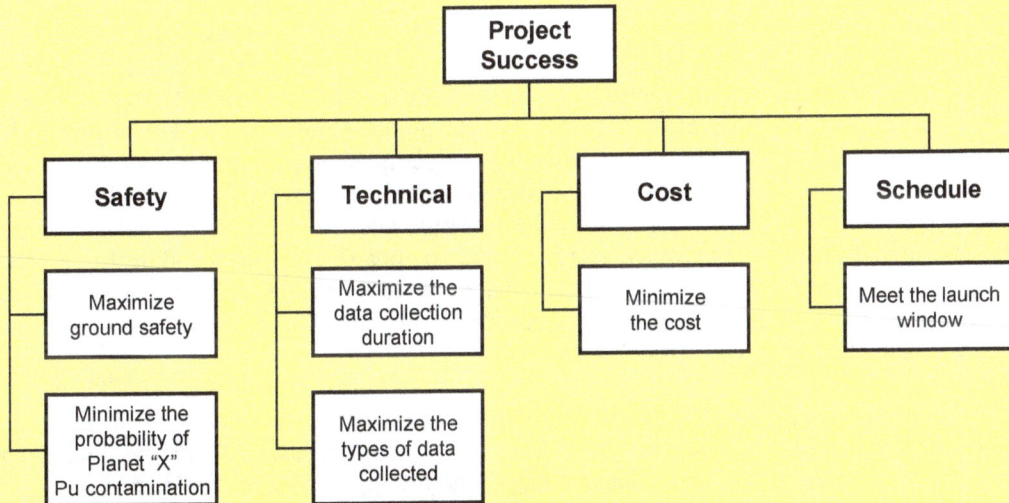

Objectives Hierarchy for the Planetary Science Mission Example

A quantitative performance measure is associated with each performance objective, along with any applicable imposed constraints. Below are the performance measures and applicable imposed constraints for four of the performance objectives. These are the four performance measures that will be quantified for the example in subsequent steps. In practice, all performance objectives are quantified.

Selected Performance Measures and Imposed Constraints
for the Planetary Science Mission Example

Performance Objective	Performance Measure	Imposed Constraint
Minimize cost	Project cost ($M)	None
Minimize development time	Months to completion	55 months
Minimize the probability of Planet "X" Pu contamination	Probability of Planet "X" Pu contamination	0.1%
Maximize data collection	Months of data collection	6 months

3.1.2   Step 2 - Compile Feasible Alternatives

The objective of Step 2 is to compile a comprehensive list of feasible decision alternatives through a discussion of a reasonable range of alternatives. The result is a set of alternatives that can potentially achieve objectives and warrant the investment of resources required to analyze them further.

3.1.2.1 Compiling an Initial Set of Alternatives

Decision alternatives developed under the design solution definition process [2] are the starting point. These may be revised, and unacceptable alternatives removed after deliberation by stakeholders based upon criteria such as violation of flight rules, violation of safety standards, etc. Any listing of alternatives will by its nature produce both practical and impractical alternatives. It would be of little use to seriously consider an alternative that cannot be adopted; nevertheless, the initial set of proposed alternatives should be conservatively broad in order to reduce the possibility of excluding potentially attractive alternatives from the outset. Keep in mind that novel solutions may provide a basis for the granting of exceptions and/or waivers from deterministic standards, if it can be shown that the intents of the standards are met, with confidence, by other means. In general, it is important to avoid limiting the range of proposed alternatives based on prejudgments or biases.

Defining feasible alternatives requires an understanding of the technologies available, or potentially available, at the time the system is needed. Each alternative should be documented qualitatively in a description sheet. The format of the description sheet should, at a minimum, clarify the allocation of required functions to that alternative's lower-level components. The discussion should also include alternatives which are capable of avoiding or substantially lessening any significant risks, even if these alternatives would be more costly. If an alternative would cause one or more significant risk(s) in addition to those already identified, the significant effects of the alternative should be discussed as part of the identification process.

Stakeholder involvement is necessary when compiling decision alternatives, to assure that legitimate ideas are considered and that no stakeholder feels unduly disenfranchised from the decision process. It is expected that interested parties will have their own ideas about what constitutes an optimal solution, so care should be taken to actively solicit input. However, the initial set of alternatives need not consider those that are purely speculative. The alternatives should be limited to those that are potentially fruitful.

3.1.2.2 Structuring Possible Alternatives (e.g., Trade Trees)

One way to represent decision alternatives under consideration is by a trade tree. Initially, the trade tree contains a number of high-level decision alternatives representing high-level differences in the strategies used to address objectives. The tree is then developed in greater detail by determining a general category of options that are applicable to each strategy. Trade tree development continues iteratively until the leaves of the tree contain alternatives that are well enough defined to allow quantitative evaluation via risk analysis (see Section 3.2).

Along the way, branches of the trade tree containing unattractive categories can be pruned, as it becomes evident that the alternatives contained therein are either *infeasible* (i.e., they are incapable of satisfying the imposed constraints) or categorically inferior to alternatives on other branches. An alternative that is inferior to some other alternative with respect to every performance measure is said to be *dominated* by the superior alternative. At this point in the RIDM process, assessment of performance is high-level, depending on simplified analysis and/or expert opinion, etc. When performance measure values are quantified, they are done so as point estimates, using a conservative approach to estimation in order to err on the side of inclusion rather than elimination.

Figure 17 presents an example launch vehicle trade tree from the Exploration Systems Architecture Study (ESAS) [20]. At each node of the tree the alternatives were evaluated for feasibility within the cost and schedule constraints of the study's ground rules and assumptions. Infeasible options were pruned (shown in red), focusing further analytical attention on the retained branches (shown in green). The key output of this step is a set of alternatives deemed to be worth the effort of analyzing with care. Alternatives in this set have two key properties:

- They do not violate imposed constraints

- They are not known to be dominated by other alternatives (i.e., there is no other alternative in the set that is superior in every way).

Alternatives found to violate either of these properties can be screened out.

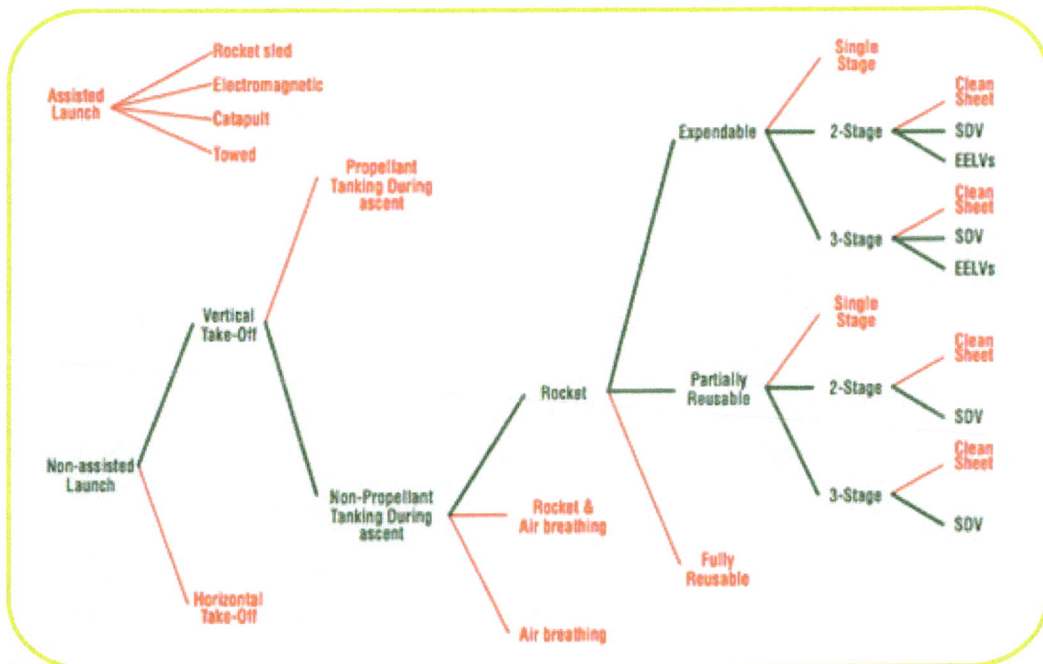

Figure 17. Example Launch Vehicle Trade Tree from ESAS

Planetary Science Mission Example: Compile Feasible Alternatives

A trade tree approach was used to develop potential alternatives for the mission to Planet "X". As shown in the trade tree below, the three attributes that were traded were the orbital insertion method (propulsive deceleration vs. aerocapture), the science package (lighter, low-fidelity instrumentation vs. heavier, high-fidelity instrumentation), and the launch vehicle (small, medium, and large). However, initial estimates of payload mass indicated that there was only one appropriately matched launch vehicle option to each combination of insertion method and science package. Thus, eight of the twelve initial options were screened out as being "infeasible" (as indicated by the red X's), leaving four alternatives to be forwarded to risk analysis (alternatives 1 – 4).

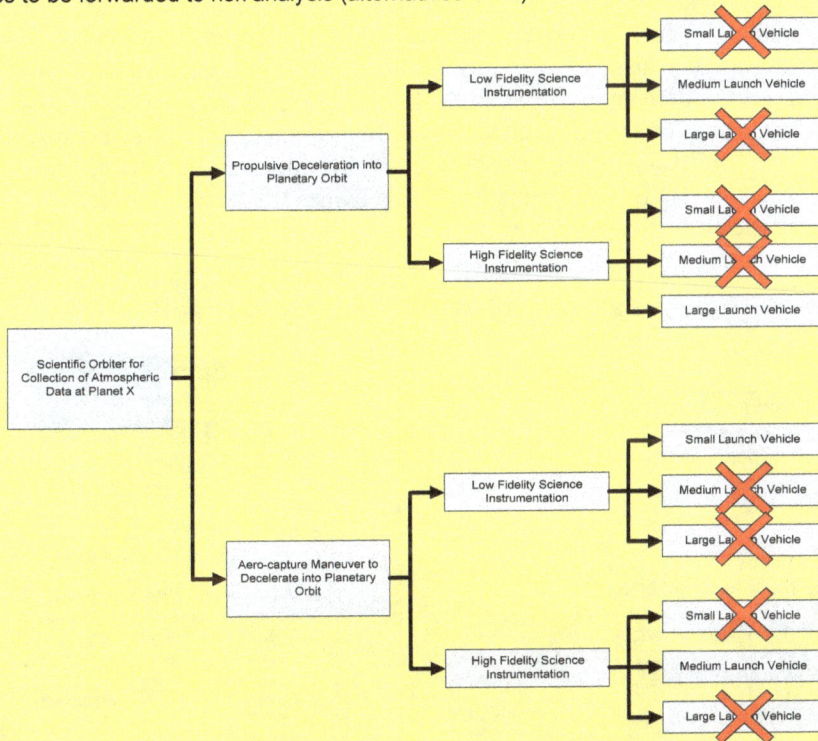

Trade Tree of Planetary Science Mission Alternatives

Feasible Alternatives forwarded to Risk Analysis

Alt #	Orbital Insertion Technology	Science Package	Launch Vehicle Size
1	Propulsive Insertion	Low Fidelity	Medium
2	Propulsive Insertion	High Fidelity	Large
3	Aerocapture	Low Fidelity	Small
4	Aerocapture	High Fidelity	Medium

3.2 Part 2 – Risk Analysis of Alternatives

Risk analysis consists of performance assessment supported by probabilistic modeling. It links the uncertainties inherent in a particular decision alternative to uncertainty in the achievement of objectives, were that decision alternative to be pursued. Performance is assessed in terms of the performance objectives developed in Step 1. The performance measures established for these objectives provide the means of quantifying performance so that alternatives can be effectively compared.

Figure 18 illustrates Part 2 of the RIDM process, Risk Analysis of Alternatives. In Step 3, risk analysis methodologies are selected for each analysis domain represented in the objectives, and coordination among the analysis activities is established to ensure a consistent, integrated evaluation of each alternative. In Step 4, the risk analysis is conducted, which entails probabilistic evaluation of each alternative's performance measure values, iterating the analysis at higher levels of resolution as needed to clearly distinguish performance among the alternatives. Then the TBfD is developed, which provides the primary means of risk-informing the subsequent selection process.

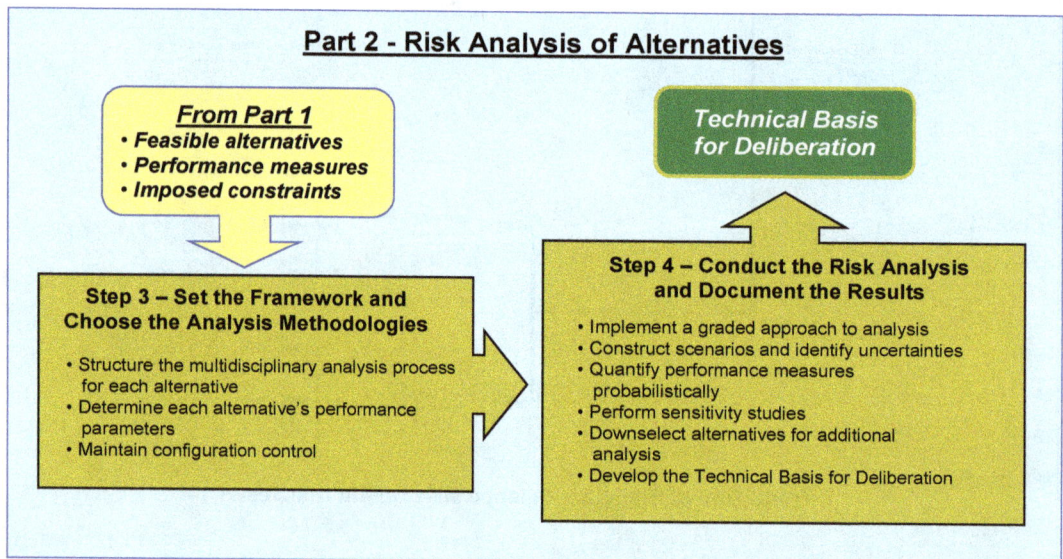

Figure 18. RIDM Process Part 2, Risk Analysis of Alternatives

3.2.1 Step 3 – Set the Framework and Choose the Analysis Methodologies

This step of the RIDM process is concerned with how domain-specific analyses, conducted in accordance with existing methodological practices, are integrated into a multidisciplinary framework to support decision making under uncertainty. In general, each mission execution domain has a suite of analysis methodologies available to it that range in cost, complexity, and time to execute, and which produce results that vary from highly uncertain rough order-of-magnitude (ROM) estimates to the detailed simulations. The challenge for the risk analysts is to establish a framework for analysis across mission execution domains that:

- Operates on a common set of (potentially uncertain) *performance parameters* for a given alternative (e.g., the cost model uses the same mass data as the lift capacity model);

- Consistently addresses uncertainties across mission execution domains and across alternatives (e.g., budget uncertainties, meteorological variability);

- Preserves correlations between performance measures (discussed further in Section 3.2.2); and

- Is transparent and traceable.

The means by which a given level of performance will be achieved is alternative specific, and accordingly, the analyses that are required to support quantification are also alternative specific. For example, one alternative might meet the objective of *Minimize Crew Fatalities* by developing a high reliability system with high margins and liberal use of redundancy, eliminating the need for an abort capability. Since the high mass associated with the high margins of this approach impacts the objective, *Maximize Payload Capacity*, a different alternative might address the same crew safety objective by combining a lighter, less reliable system with an effective crew abort capability. For these two alternatives, significantly different analyses would need to be performed to quantify the probability *P(LOC)* of accomplishing the crew safety performance measure. In the first case, *P(LOC)* is directly related to system reliability. In the second case, reliability analysis plays a significant part, but additional analysis is needed to quantify abort effectiveness, which involves analysis of system responsiveness to the failure, and survivability given the failure environment.

Performance Parameters

A *performance parameter* is any value needed to execute the models that quantify the performance measures. Unlike performance measures, which are the same for all alternatives, performance parameters typically vary among alternatives, i.e., a performance parameter that is defined for one alternative might not apply to another alternative.

Example performance parameters related to the performance objective of lofting X lbs into low Earth orbit (LEO) might include propellant type, propellant mass, engine type/specifications, throttle level, etc. Additionally, performance parameters also include relevant environmental characteristics such as meteorological conditions.

Performance parameters may be uncertain. Indeed, risk has its origins in performance parameter uncertainty, which propagates through the risk analysis, resulting in performance measure uncertainty.

3.2.1.1 Structuring the Analysis Process

For a given alternative, the relationship between performance measures and the analyses needed to quantify them can be established and illustrated using a means objectives network (introduced in Section 3.1.1). Figure 19, adapted from [21], illustrates the idea. This figure traces Performance Parameter 1 through the risk analysis framework, showing how it is used by multiple risk analyses in multiple mission execution domains.

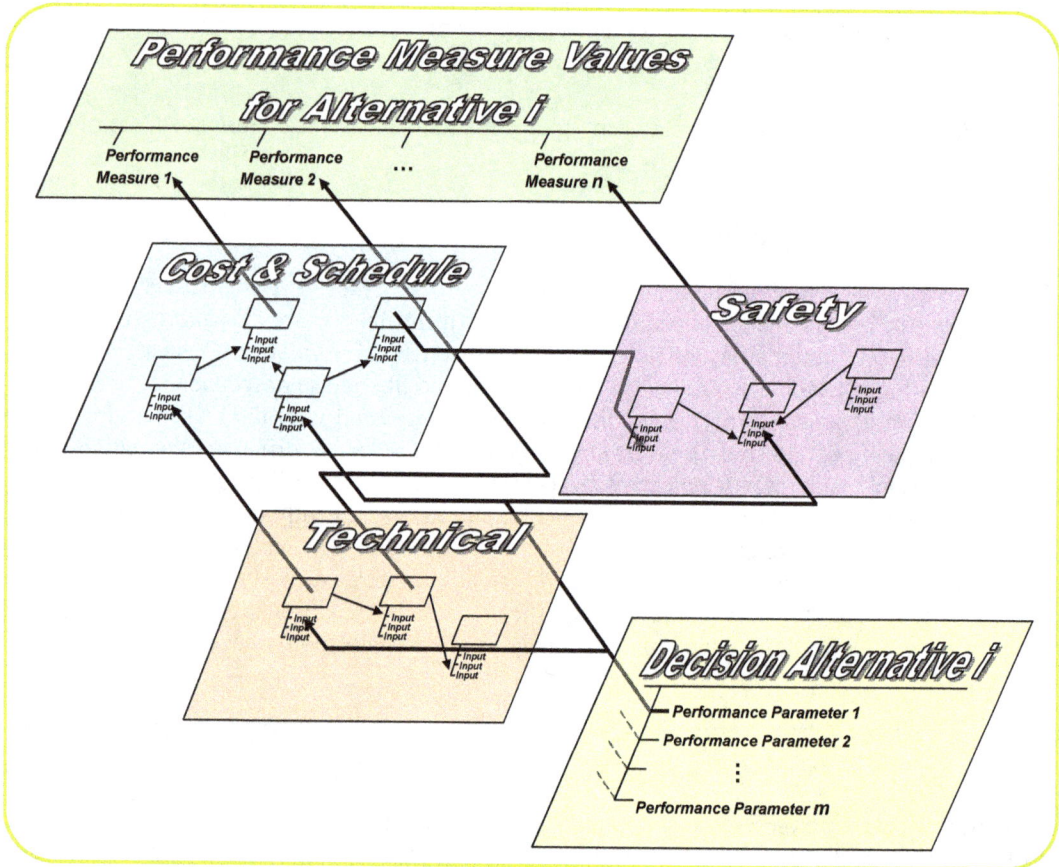

Figure 19. Risk Analysis Framework (Alternative Specific)

For example, Performance Parameter 1 is a direct input to a risk analysis in the Cost and Schedule mission execution domains (which have been combined in the figure for convenience). This analysis produces outputs that are used as inputs to two other Cost and Schedule risk analyses. One of these risk analyses produces a value for Performance Measure 1, whereas the other risk analysis produces an output that is needed by a risk analysis in the Safety mission execution domain. This Safety risk analysis ultimately supports quantification of Performance Measure **n**.

Each of the **m** performance parameters that defines Alternative *i* can be similarly traced through the risk analysis framework.

Figure 19 illustrates the need for coordination among the organizations conducting the analyses to assure that:

- There is an organization responsible for the quantification of each performance measure;

- The data requirements for every risk analysis are understood and the data sources and destinations have been identified;

- All data are traceable back through the risk analysis framework to the performance parameters of the analyzed alternative.

3.2.1.2 Configuration Control

It is important to maintain consistency over the definition of each analyzed alternative to ensure that all involved parties are working from a common data set. This is particularly true during the earlier phases of the program/project life cycle where designs may be evolving rapidly as decisions are made that narrow the trade space and extend it to higher levels of detail. It is also true when decisions are revisited, such as during requirements rebaselining (as discussed in Section 2.2), in which case the complete definition of the alternative may be distributed among various organizational units at different levels of the NASA hierarchy. In this case it is necessary for the organization at the level of the decision to be made to consolidate all relevant alternative data at its own level, as well as levels below, into a configuration managed data set.

Additionally, the risk analysis framework itself must be configuration controlled, in terms of the analyses (e.g., version number) and data pathways.

3.2.1.3 Implementing a Graded Approach in Selecting Risk Analysis Methods

The spectrum of analysis disciplines involved in the risk analysis of alternatives is as broad as the spectrum of performance measures, spanning the mission execution domains of Safety, Technical, Cost, and Schedule. It is not the intent of this handbook to provide detailed guidance on the conduct of domain-specific analyses. Such guidance is available in domain-specific documents like NPR 8715.3C, NASA General Safety Program Requirements [22] (Chapter 2, System Safety), the NASA Cost Estimating Handbook [23], the NASA Systems Engineering Handbook [2], and the NASA Probabilistic Risk Assessment Procedures Guide [24].

Depending on project scale, life cycle phase, etc., different levels of analysis are appropriate. The rigor of analysis should be enough to assess compliance with imposed constraints and support selection between alternatives. Iteration is to be expected as part of the analysis process, but as a general rule of thumb, the rigor of analysis should increase with successive program/project life cycle phases. In addition for a given phase, parametric, engineering, and logic modeling can commence at a low level of detail; the level of detail can be increased in an iterative fashion based on the requirement to reach a robust decision. Figure 20 indicates the types of analysis that are generally appropriate, as a function of life cycle phase, for cost, technical, and safety estimation. Discussion of uncertainty can be found in Section 3.2.2. Detailed information on methods can be found in discipline-specific guidance, e.g., [2], [23], and [24].

Cost Estimating Methodology Guidance Chart

	Pre-Phase A	Phase A	Phase B	Phase C/D	Phase E
Analogy	●	◐	◐	◐	○
Parametric	●	●	◐	◐	○
Engineering Build Up	◐	◐	●	●	●

Technical Estimating Methodology Guidance Chart

	Pre-Phase A	Phase A	Phase B	Phase C/D	Phase E
First-Order	●	●	◐	○	○
Detailed Simulation	○	●	●	●	◐
Testing	○	○	◐	●	●
Operating Experience	○	○	○	○	●

Safety, Reliability, and Operations Estimating Methodology Guidance Chart

	Pre-Phase A	Phase A	Phase B	Phase C/D	Phase E
Similarity	●	◐	◐	○	○
First-Order Parametric	●	●	◐	◐	○
Detailed Logic Modeling	○	○	●	●	●
Statistical Methods	○	○	○	◐	●

Legend: ● Primary ◐ Applicable ○ Typically Not Applicable

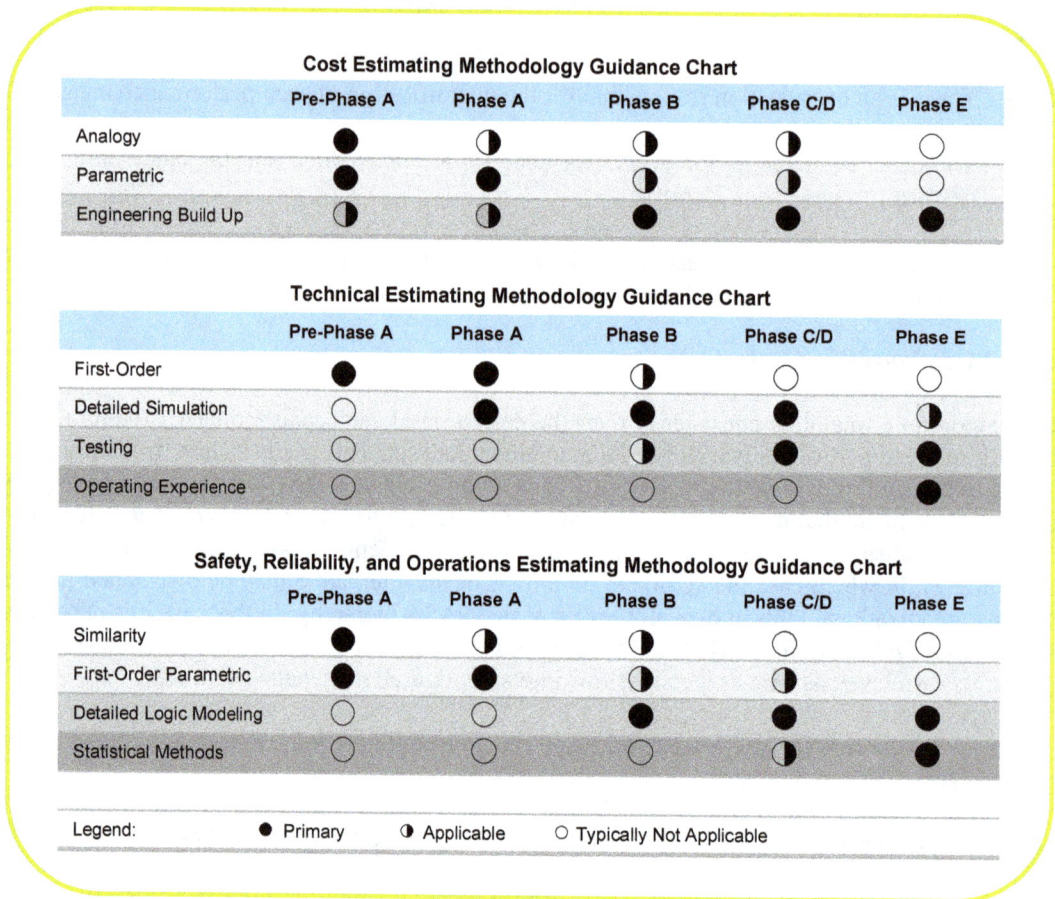

Figure 20. Analysis Methodology Guidance Chart

- **Cost and Schedule Estimating Methodologies:**

  o *Analogy Estimating Methodology* - Analogy estimates are performed on the basis of comparison and extrapolation to like items or efforts. Cost data from one past program that is technically representative of the program to be estimated serves as the basis of estimate. These data are then subjectively adjusted upward or downward, depending upon whether the subject system is believed to be more or less complex than the analogous program.

  o *Parametric Estimating* - Estimates created using a parametric approach are based on historical data and mathematical expressions relating cost as the dependent variable to selected, independent, cost-driving variables through regression analysis. Generally, an estimator selects parametric estimating when only a few key pieces of data are known, such as weight and volume. The implicit assumption of parametric estimating is that the same forces that affected cost in the past will affect cost in the future.

- o *Engineering Build Up Methodology* - Sometimes referred to as "grass roots" or "bottom-up" estimating, the engineering build up methodology rolls up individual estimates for each element into the overall estimate. This methodology involves the computation of the cost of a work breakdown structure (WBS) element by estimating at the lowest level of detail (often referred to as the "work package" level) wherein the resources to accomplish the work effort arc readily distinguishable and discernable. Often the labor requirements are estimated separately from material requirements. Overhead factors for cost elements such as Other Direct Costs (ODCs), General and Administrative (G&A) expenses, materials burden, and fees are generally applied to the labor and materials costs to complete the estimate.

- **Estimating Methodologies for Technical Performance Measures:**

  - o *First-Order Estimating Methodology* - First-order estimates involve the use of closed-form or simple differential equations which can be solved given appropriate bounding conditions and/or a desired outcome without the need for control-volume based computational methods. The equations may be standard physics equations of state or empirically-derived relationships from operation of similar systems or components.

  - o *Detailed Simulation Estimating Methodology* - Estimates using a detailed simulation require the construction of a model that represents the physical states of interest in a virtual manner using control-volume based computational methods or methods of a similar nature. These simulations typically require systems and conditions to be modeled to a high-level of fidelity and the use of "meshes" or network diagrams to represent the system, its environment (either internal, external, or both), and/or processes acting on the system or environment. Examples are computational fluid dynamics (CFD) and finite-element modeling.

  - o *Testing Methodology* - Testing can encompass the use of table-top experiments all the way up to full-scale prototypes operated under real-world conditions. The objective of the test is to measure how the system or its constituent components may perform within actual mission conditions. Testing could be used for assessing the expected performance of competing concepts or for evaluating that the system or components will meet flight specifications.

  - o *Operating Experience Methodology* - Once the system is deployed data gathered during operation can be analyzed to provide empirically accurate representations of how the system will respond to different conditions and how it will operate throughout its lifetime. This information can serve as the basis for applicable changes, such as software uploads or procedural changes, that may improve the overall performance of the system. Testing and detailed simulation may be combined with operating experience to extrapolate from known operating conditions.

- **Safety, Reliability, & Operations Estimating Methodologies:**

  o *Similarity Estimating Methodology* - Similarity estimates are performed on the basis of comparison and extrapolation to like items or efforts. Reliability and operational data from one past program that is technically representative of the program to be estimated serves as the basis of estimate. Reliability and operational data are then subjectively adjusted upward or downward, depending upon whether the subject system is believed to be more or less complex than the analogous program.

  o *First-Order Parametric Estimation* - Estimates created using a parametric approach are based on historical data and mathematical expressions relating safety, reliability, and/or operational estimates as the dependent variable to selected, independent, driving variables through either regression analysis or first-order technical equations (e.g., higher pressures increase the likelihood of tank rupture). Generally, an estimator selects parametric estimating when the system and its concept of operation are at the conceptual stage. The implicit assumption of parametric estimating is that the same factors that shaped the safety, reliability, and operability in the past will affect the system/components being assessed.

  o *Detailed Logic Modeling Estimation* - Detailed logic modeling estimation involves "top-down" developed but "bottom-up" quantified scenario-based or discrete-event logic models that segregate the system or processes to be evaluated into discrete segments that are then quantified and mathematically integrated through Boolean logic to produce the top-level safety, reliability, or operational estimate. Detailed technical simulation and/or testing, as well as operational data, can be used to assist in developing pdfs for quantification of the model. Typical methods for developing such models may include the use of fault trees, influence diagrams, and/or event trees.

  o *Statistical Methods* - Statistical methods can applied to data collected during system/component testing or from system operation during an actual mission. This is useful for characterizing the demonstrated safety, reliability, or operability of the system. In addition, patterns in the data may be modeled in a way that accounts for randomness and uncertainty in the observations, and then serve as the basis for design or procedural changes that may improve the overall safety, reliability, or operability of the system. These methods are useful for answering yes/no questions about the data (hypothesis testing), describing associations within the data (correlation), modeling relationships within the data (regression), extrapolation, interpolation, or simply for data mining activities.

3.2.1.4 Implementing a Graded Approach in Quantifying Individual Scenarios

In addition to increasing with successive program/project life cycle phases and the level of design detail available, the level of rigor in the analysis should increase with the importance of the scenario being evaluated. Regardless of the time during the life cycle, certain scenarios will not be as important as others in affecting the performance measures that can be achieved for a given alternative. Scenarios that can be shown to have very low likelihood of occurrence and/or very low impacts on all the mission execution domains do not have to be evaluated using a rigorous simulation methodology or a full-blown accounting of the uncertainties. A point-estimate analysis using reasonably conservative simulation models and input parameter values should be sufficient for the evaluation of such scenarios.

3.2.1.5 Use of Existing Analyses

The RIDM process does not imply a need for a whole new set of analyses. In general, some of the necessary analyses will already be planned or implemented as part of the systems engineering, cost estimating, and safety and mission assurance (S&MA) activities. Risk analysis for RIDM should take maximum advantage of existing activities, while also influencing them as needed in order to produce results that address objectives, at an appropriate level of rigor to support robust decision making.

Planetary Science Mission Example: Set the Analysis Framework

The figure below shows the risk analysis framework used to integrate the domain-specific analyses. Each alternative is characterized by its performance parameters, some of which are uncertain (shown in red text) and others of which have definite, known deterministic values (shown in black text). In order to calculate the performance measures previously selected for illustrative purposes, four separate performance models have been developed for radiological contamination, data collection, schedule, and cost. Some performance parameters, such as *spacecraft structure mass*, *launch reliability*, and *science package TRL*, are used in multiple models. Some models (e.g., the data collection model) produce outputs (e.g., *science package mass*) that are inputs to other models (e.g., the schedule model).

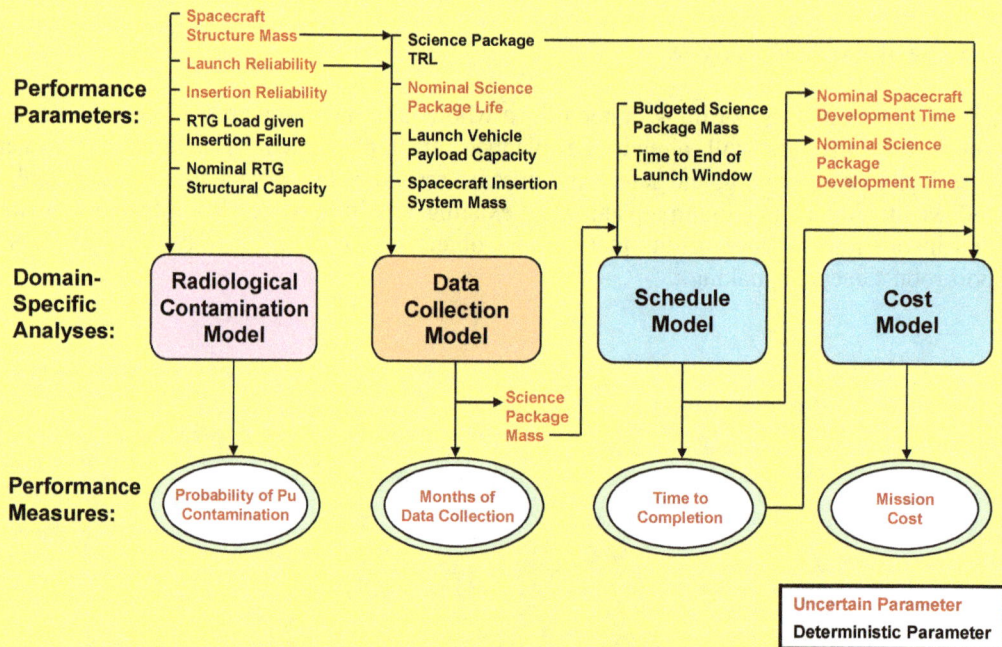

Risk Analysis Framework

The analysis framework shown above was used for all four alternatives selected for risk analysis. However, in general, each alternative may require its own analysis framework, which may differ substantially from other alternatives' frameworks based on physical or operational differences in the alternatives themselves. When this is the case, care should be taken to assure analytical consistency among alternatives in order to support valid comparisons.

3.2.2 Step 4 – Conduct the Risk Analysis and Document the Results

Once the risk analysis framework is established and risk analysis methods determined, performance measures can be quantified. As discussed previously, however, this is just the start of an iterative process of successive analysis refinement driven by stakeholder and decision-maker needs (see Part 3 of the RIDM process).

3.2.2.1 Probabilistic Modeling of Performance

If there were no uncertainty, the question of performance assessment would be one of quantifying point value performance measures for each decision alternative. In the real world, however, uncertainty is unavoidable, and the consequences of selecting a particular decision alternative cannot be known with absolute precision. When the decision involves a course of action there is uncertainty in the unfolding of events, however well planned, that can affect the achievement of objectives. Budgets can shift, overruns can occur, technology development activities can encounter unforeseen phenomena (and often do). Even when the outcome is realized, uncertainty will still remain. Reliability and safety cannot be known absolutely, given finite testing and operational experience. The limits of phenomenological variability in system performance can likewise not be known absolutely nor can the range of conditions under which a system will have to operate. All this is especially true at NASA, which operates on the cutting edge of scientific understanding and technological capability.

For decision making under uncertainty, risk analysis is necessary, in which uncertainties in the values of each alternative's performance parameters are identified and propagated through the analysis to produce uncertain performance measures (see Figure 6 in Section 1.5). Moreover, since performance measures might not be independent, correlation must be considered. For example, given that labor tends to constitute a high fraction of the overall cost of many NASA activities, cost and schedule tend to be highly correlated. High costs tend to be associated with slipped schedules, whereas lower costs tend to be associated with on-time execution of the program/project plan.

One way to preserve correlations is to conduct all analysis within a common Monte Carlo "shell" that samples from the common set of uncertain performance parameters, propagates them through the suite of analyses, and collects the resulting performance measures as a vector of performance measure values [25]. As the Monte Carlo shell iterates, these performance measure vectors accumulate in accordance with the parent joint pdf that is defined over the entire set of performance measures. Figure 21 notionally illustrates the Monte Carlo sampling procedure as it would be applied to a single decision alternative (Decision Alternative i).

Uncertainties are distinguished by two categorical groups: aleatory and epistemic [26], [27]. Aleatory uncertainties are random or stochastic in nature and cannot be reduced by obtaining more knowledge through testing or analysis. Examples include:

- The room-temperature properties of the materials used in a specific vehicle.

- The scenario(s) that will occur on a particular flight.

In the first case, there is random variability caused by the fact that two different material samples will not have the same exact properties even though they are fabricated in the same manner. In the second case, knowing the mean failure rates for all the components with a high degree of certainty will not tell us which random failures, if any, will actually occur during a particular flight. On the other hand, epistemic uncertainties are not random in nature and can be reduced by obtaining more knowledge through testing and analysis. Examples include:

- The properties of a material at very high temperatures and pressures that are beyond the capability of an experimental apparatus to simulate.

- The mean failure rates of new-technology components that have not been exhaustively tested to the point of failure in flight environments.

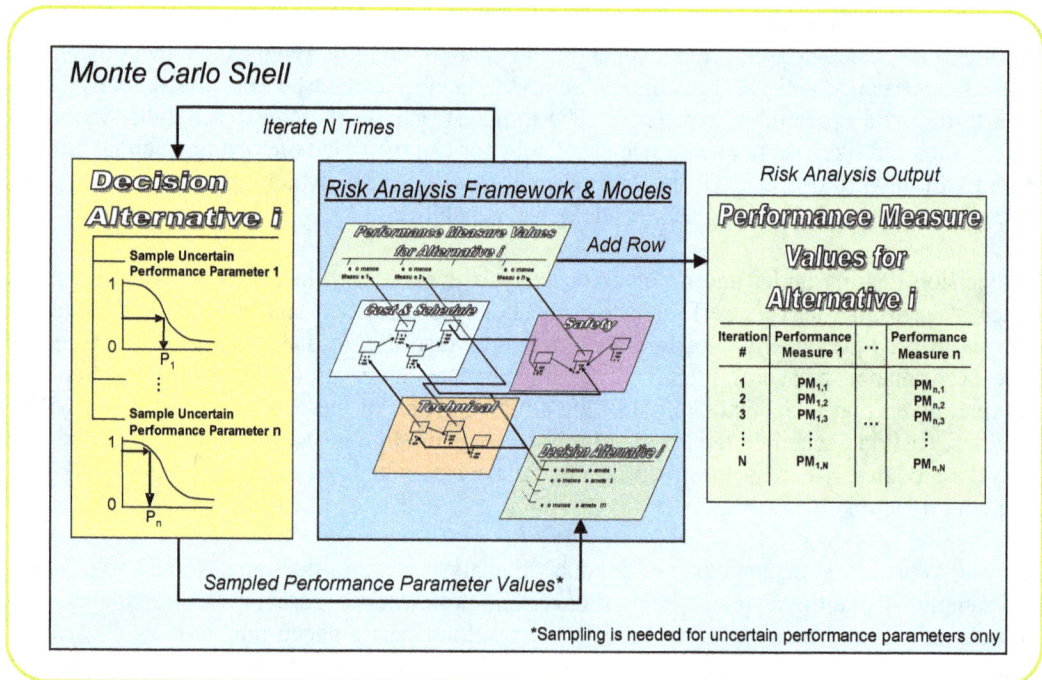

Figure 21. Risk Analysis Using a Monte Carlo Sampling Procedure

In both cases, the uncertainty is caused by missing or incomplete knowledge or by limitations in the models used to make predictions.

The assessed performance of an alternative is affected by both types of uncertainty in basically the same way. That is to say, if there were no epistemic uncertainties there would still be uncertainty in the assessed performance because of the unpredictability of performance parameters that are subject to aleatory uncertainty. Likewise, if there were no aleatory uncertainties there would still be uncertainty in the assessed performance caused by the possibility of our mischaracterizing reality due to imperfect information.

It has become common in risk analysis to separate these two contributions to uncertainty by using the term *risk* to reflect the variability caused by aleatory uncertainties alone, and the term *risk uncertainty* or simply *uncertainty* to reflect the impreciseness of our knowledge of the risk caused by epistemic uncertainties alone. This distinction is useful for deciding whether additional research is worth the cost that it would entail, but is not as important for distinguishing between different architectural or design alternatives. Therefore, for purposes of the RIDM process, we speak only of uncertainties in the broad sense and do not distinguish between their aleatory and epistemic parts. However, the analyst always has the option of keeping aleatory and epistemic uncertainties separate from one another if he or she desires to do so, and in CRM where mitigation options are considered, this separation can be essential.

Further arguments about the relative advantages of combining aleatory and epistemic uncertainties versus keeping them separate may be found in [28].

3.2.2.2 Use of Qualitative Information in RIDM

As discussed in the preceding section, uncertainties in the forecasted performance measures are caused by uncertainties in the input performance parameters and in the models that are used to calculate the outcomes. These parameter and modeling uncertainties may be expressed in either quantitative or qualitative terms. If a parameter is fundamentally quantitative in nature, it is represented as having an uncertainty distribution that is expressed in terms of numerical values. For example, the date that a part is delivered is a quantitative performance parameter because it is defined in terms of the number of days between a reference date (e.g., the project's initiation) and the delivery date. The date has a discrete numerical distribution because it changes in 24-hour increments. Most performance parameters, such as the cost of the part or its failure rate, have continuous numerical distributions.

A performance parameter can often also be expressed in terms of a constructed scale that is qualitative in nature. For example, the technology readiness level (TRL) at the time of project initiation is a qualitative parameter because it is defined in terms of ranks that are based on non-numerical information. A TRL of 1, for example, is defined by terms such as: "basic principles observed and reported," "transition from scientific research to applied research," "essential characteristics and behaviors of systems and architectures," "descriptive tools are mathematical formulations or algorithms." Such terms are not amenable to quantitative analysis without a significant amount of interpretation on the part of the analysts.

While the performance parameter may be either quantitative or qualitative, the probability scale for the uncertainty distribution of the performance parameter is generally defined in a quantitative manner. The probability scale may be either continuous or discrete (although in most cases it is continuous). For example, a five-tiered discretization of probabilities on a logarithmic scale might be based on binning the probabilities into the following ranges: 10^{-5} to 10^{-4} for level 1, 10^{-4} to 10^{-3} for level 2, 10^{-3} to 10^{-2} for level 3, 10^{-2} to 10^{-1} for level 4, and 10^{-1} to 100 for level 5. It could be argued that the probability levels could also be defined in verbal terms such as "very unlikely to happen," "moderately likely to happen," and "very likely to happen." While these definitions are not numerical as stated, it is usually possible to ascertain the numerical ranges that the analyst has in mind when making these assignments. Thus, the probability should be relatable to a quantitative scale.

Various types of quantitative and qualitative uncertainty distributions for the input parameters and conditions are shown in Figure 22. Three of these (the top left and right charts and the lower right chart within the first bracket) are types of probability density functions, whereas the fourth chart (lower left) is a form of a complementary cumulative distribution function (CCDF). Either form of distribution (density form or cumulative form) may be used to express uncertainty. The choice is governed by whichever is the easier to construct, based on the content of the uncertainty information.

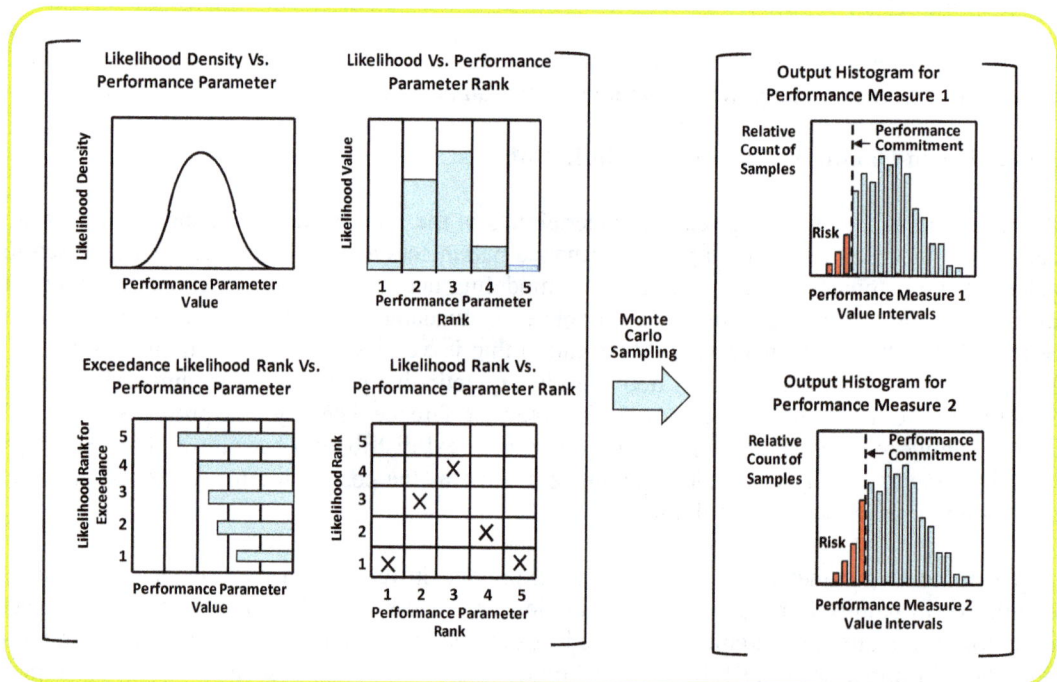

Figure 22. Uncertain Performance Parameters Leading to Performance Measure Histograms

As depicted in Figure 22, the values of the output performance measures, as opposed to the values of the input performance parameters, are always quantitative in that they are defined in terms of numerical metrics. The output uncertainty distributions are expressed in the form of a histogram representation of output values obtained from Monte Carlo sampling of the input values and conditions.

Because the numerically based models are set up to accept numerical inputs, execution of the models for calculating the output performance measures is in general easier if all the performance parameters are defined in terms of quantitative scales, whether continuous or discrete. Caution should be used where one or more of the inputs are defined in terms of a qualitative, or constructed, scale. In these cases, the calculation of the performance measures may require that different models be used depending on the rank of the qualitative input. For example, the initial TRL for an engine might depend upon whether it can be made out of aluminum or has to be made out of beryllium. In this case, an aluminum engine has a higher TRL than a beryllium engine because the former is considered a heritage engine and the latter a developmental engine. On the other hand, a beryllium engine has the potential for higher thrust because it can run at higher temperatures. The model for calculating performance measures such as engine start-up reliability, peak thrust, launch date, and project cost would likely be different for an aluminum engine than for a beryllium engine.

3.2.2.3 Risk Analysis Support of Robust Decision Making

Because the purpose of risk analysis in RIDM is to support decision making, the adequacy of the analysis methods must be determined in that context. The goal is a robust decision, where the decision-maker is confident that the selected decision alternative is actually the best one, given the state of knowledge at the time. This requires the risk analysis to be rigorous enough to discriminate between alternatives, especially for those performance measures that are determinative to the decision.

Figure 23 illustrates two hypothetical situations, both of which involve a decision situation having just one performance measure of significance. The graph on the left side of the figure shows a situation where Alternative 2 is clearly better than Alternative 1 (assuming that the pdfs are independent of each other) because the bulk of its pdf is to the left of Alternative 1's pdf. Thus the decision to select Alternative 2 is robust because there is high probability that a random sample from Alternative 1's pdf would perform better than a random sample from Alternative 2's pdf. In contrast, the graph on the right side of the figure shows a situation where the mean value of Alternative 1's performance measure is better than the mean value of Alternative 2's, but their pdfs overlap to a degree that prevents the decision to select Alternative 1 from being robust; that is, unless the pdfs for Alternatives 1 and 2 are highly correlated, there is a significant probability that Alternative 2 is actually better. The issue of correlated pdfs will be taken up later in this section.

For decisions involving multiple objectives and performance measures, it is not always possible to identify *a priori* which measures will be determinative to the decision and which will only be marginally influential. It is possible that some performance measures would require extensive analysis in order to distinguish between alternatives, even though the distinction would ultimately not be material to the decision. Consequently, the need for additional analysis for the purpose of making such distinctions comes from the deliberators and the decision-maker, as they deliberate the merits and drawbacks of the alternatives. The judgment of whether uncertainty reduction would clarify a distinction between contending decision alternatives is theirs to make; if it would be beneficial and if additional analysis is practical and effective towards that purpose, then the risk analysis is iterated and the results are updated accordingly.

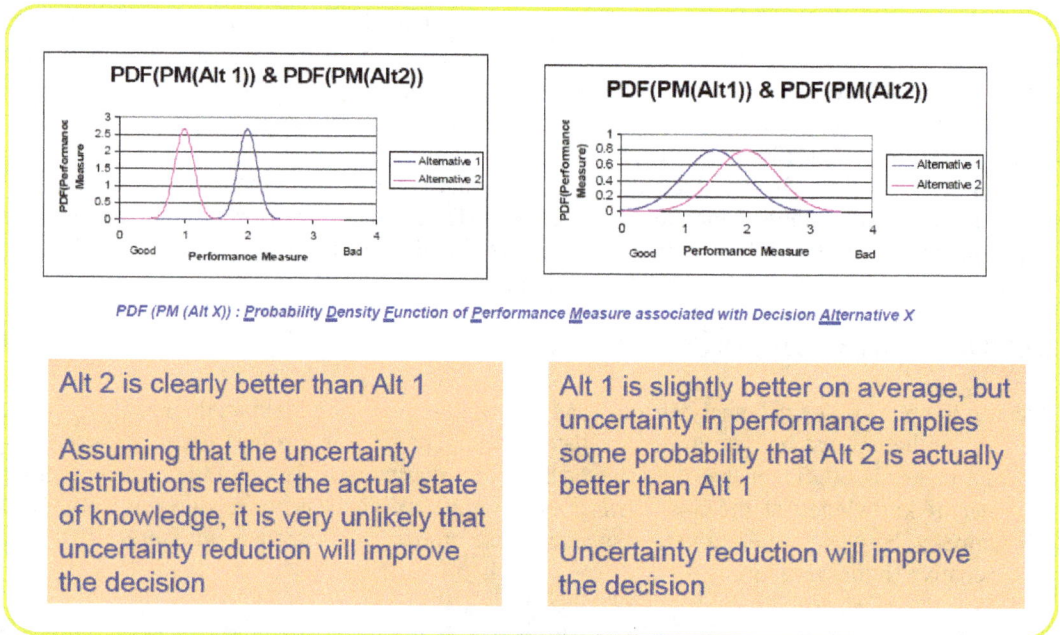

Figure 23. Robustness and Uncertainty

3.2.2.4 Sequential Analysis and Downselection

While the ultimate selection of any given alternative rests squarely with the decision maker, he or she may delegate preliminary downselection authority to a local proxy decision-maker, in order to reduce the number of contending alternatives as early as practical in the decision-making process. There is no formula for downselection; it is an art whose practice benefits from experience. In general it is prudent to continuously screen the alternatives throughout the process. It is important to document the basis for eliminating such alternatives from further consideration at the time they are eliminated. Two such bases that were discussed in Section 3.1.2 are infeasibility and dominance. Additional discussion of downselection is presented in Section 3.3.2.

Downselection often involves the conduct of sequential analyses, each of which is followed by a pruning of alternatives. In this way, alternatives that are clearly unfavorable due to their performance on one (or few) performance measures can be eliminated from further analysis once those values are quantified.

For example, a lunar transportation architecture with Earth orbit rendezvous will require some level of loiter capability for the element(s) that are launched first (excluding simultaneous-launch options). A trade tree of architecture options might include a short loiter branch and a long loiter branch, corresponding to the times needed to span different numbers of trans-lunar injection

(TLI) windows. It may not be known, prior to analysis, that the effects of propellant boil-off in terms of increased propellant needs, tankage, structure, and lift capacity, are prohibitive and render the long loiter option unfavorable. However, this circumstance can be determined based on an assessment of boil-off rate, and a sizing analysis for the architecture in the context of its DRMs. Once an analysis of sufficient rigor is performed, the entire long loiter branch of the trade tree can be pruned, without the need for additional, higher-level-of-rigor analyses on the potentially large number of long loiter alternatives compiled in Step 3 of the RIDM process.

Within the constraints of the analytical dependencies established by the risk analysis framework set in the previous step, it may be prudent to order the conduct of domain-specific analyses in a manner that exploits the potential for pruning alternatives prior to forwarding them for additional analysis. There is no hard rule for an optimal ordering; it depends on the specific decision being made, the alternatives compiled, and the analysis methods employed. It is recommended that opportunities for sequential analysis and downselection be looked for as alternatives are analyzed, and that the ordering of analyses be adjusted as appropriate to facilitate downselection, depending on which performance measures can be used as a basis for pruning (see Figure 24).

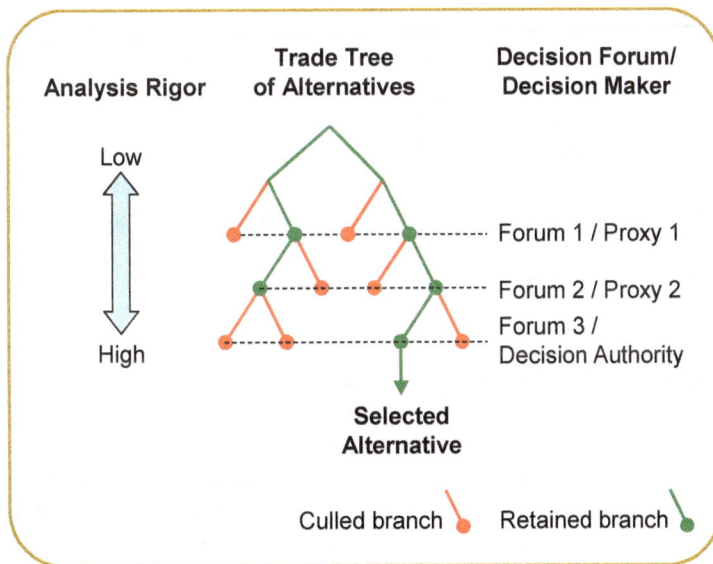

Figure 24. Downselection of Alternatives

Sequential analysis and downselection requires active collaboration among the risk analysts, the deliberators and the decision maker. It is not the role of the risk analysts to eliminate alternatives except on the grounds of infeasibility. Sequential downselection, like all decision making, must be done in the context of stakeholder values and decision-maker responsibility/accountability.

Additionally, there is a potential vulnerability to sequential downselection, due to the incomplete quantification of performance measures it entails. It assumes that for the pruned alternatives, the level of performance in the analyzed domains is so poor that no level of performance in the unanalyzed domains could possibly make up for it. If this is not actually the case, an alternative that is attractive overall might be screened out due to inferior performance in just one particular area, despite superior performance in other domains and overall. Thus, it is good practice to review the validity of the downselects, in the context of the selected alternative, in order to assure that the selected alternative is indeed dominant.

3.2.2.5 Model Uncertainty and Sensitivity Studies

As is the case with all modeling activities, risk modeling typically entails a degree of model uncertainty to the extent that there is a lack of correspondence between the model and the alternative being modeled.

Many papers have been written on the subject of characterizing and quantifying model uncertainties; for example, surveys may be found in [29] and [30]. Often, the approaches advocate the use of expert judgment to formulate uncertainty distributions for the results from the models.

For example, suppose there was an existing model that produced a point value for thrust (a performance measure) based on a correlation of experimental data. In developing the correlation, the analysts emphasized the data points that produced lower thrust values over those that produced higher values in keeping with engineering practice to seek a realistically conservative result. In addition, the correlation was further biased to lower values to account for the fact that the experiments did not duplicate the high temperature, high pressure environment that is experienced during flight. A modified model was also derived that was similar to the original model but did not include any biasing of the data to produce a conservative result.

Based on this evidence, a set of subject matter experts made the following judgments:

- There is a 95% likelihood that the thrust during an actual flight will be higher than that predicted by the first model, which is known to be conservative.

- There is a 25% likelihood that the actual thrust will be higher than what is being predicted by the modified model, because the model does not introduce conservative assumptions in the data analysis, and in addition the experimental simulation does not cover a range of environments where certain phenomena could decrease the thrust.

- There is only a 1% likelihood that the thrust will be lower than 0.8 times the values predicted by the original model, because there are no data to indicate that the thrust could be so low.

- There is a 1% likelihood that the thrust will be higher than 1.4 times the values predicted by the modified model because neither the original nor the modified model accounts for catalysis effects which could increase the thrust by up to 40%.

The analysts take this information to create a continuous distribution for the ratio of the actual thrust to that predicted by the original model (Figure 25). Thus, the modeling uncertainty for thrust is characterized by an adjustment factor that has a defined uncertainty distribution and is applied directly to the model output.

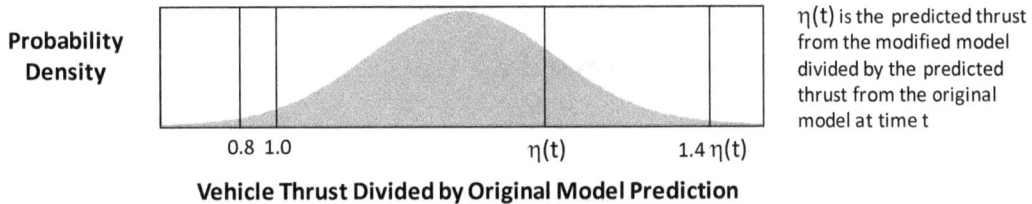

Figure 25. Conceptualization of the Formulation of Modeling Uncertainty

The technique described above, where one or more models is subjectively assessed for bias as a means of quantifying model uncertainty, can in some sense be considered part of the overall modeling effort, as opposed to being a follow-on process that is applied to the risk model results. It is generically applicable in situations where subjective expertise, above and beyond that which is already captured in the available models, can be brought to bear to construct a "meta" model that integrates the information that is available from potentially diverse sources.

Another means of assuring that decisions are robust with respect to model uncertainty is to conduct sensitivity studies over ranges of credible model forms and/or parameter values. Sensitivity studies are particularly pertinent for models that produce point value performance measure results, even when the performance measure is known to be uncertain. In these cases, it is valuable to determine the sensitivity of the decision to bounding variations in the risk model assumptions. Figure 26 notionally presents the results of such a study. It shows how the preferred alternative varies as a function of assumptions about contractor support cost rate and payload mass. For example, if the contractor support cost rate is 120 and the payload mass is 18, then Alternative A is the preferred alternative. If, however, the assumed payload mass is 4, then Alternative B is preferable. More generally, if "Alternative B" is preferred for all reasonable values of contractor support cost rate and payload mass, then the decision is robust in favor of Alternative B (with respect to these parameters), without the need for additional rigor in determining the actual contractor support cost rate or payload mass. Likewise, if the reasonable range of these parameters falls entirely within the region "Alternative A," then the decision is robust for Alternative A. Only when the reasonable range of values straddles more than one region is more rigorous characterization of contractor support cost and payload mass needed for robust decision making.

3.2.2.6 Analysis Outputs

Like the variation in risk analysis methods, the analysis results presentation for RIDM may vary, depending on the nature of the problem being evaluated. Consequently, there can be no one standard analysis output. Instead, the results are tailored to the problem and the needs of the deliberation process. Consideration should be given for providing a variety of results, including:

- Scenario descriptions

- Performance measure pdfs and statistics

- Risk results (e.g., risk of not meeting imposed constraints)

- Uncertainty analyses and sensitivity studies

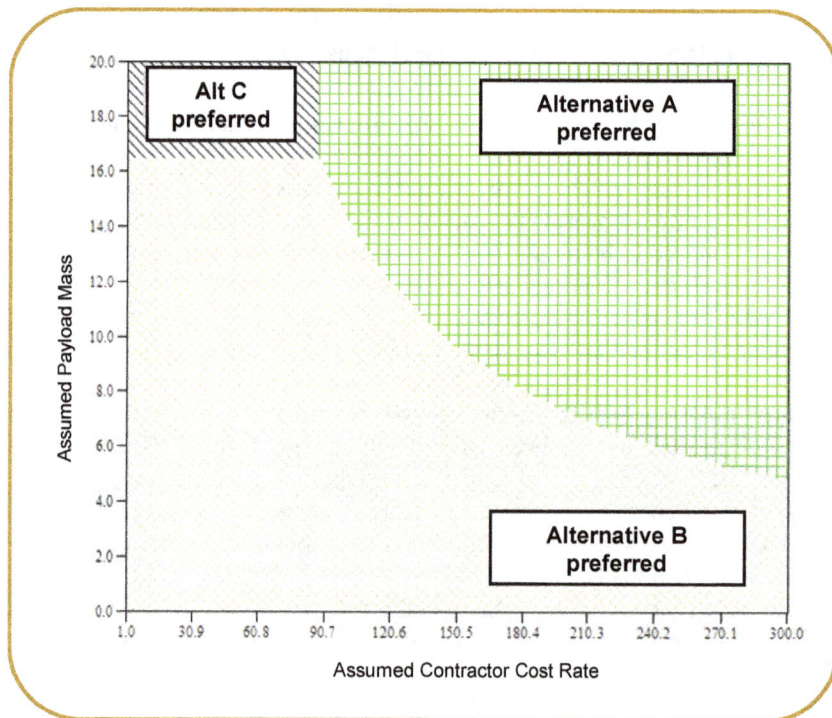

Figure 26. Notional Depiction of Decision Sensitivity to Input Parameters

It is important to note that the risk analysis results are expected to mature and evolve as the analysis iterates with the participation of the stakeholders and the decision-maker. This is not only due to increasing rigor of analysis as the stakeholders and the decision-maker strive for decision robustness. Additionally, as they establish firm performance commitments, it becomes possible to evaluate the analysis results in the context of those commitments. For example, prior to the development of performance commitments, it is not possible to construct a risk list that is keyed to the performance measures (except with respect to imposed constraints, which are firmly established prior to analysis).

3.2.2.7 Assessing the Credibility of the Risk Analysis Results

In a risk-informed decision environment, risk analysis is just one element of the decision-making process, and its influence on the decision is directly proportional to the regard in which it is held by the deliberators. A well-done risk analysis whose merits are underappreciated might not influence a decision significantly, resulting in a lost opportunity to use the available information to best advantage. Conversely, an inferior risk analysis held in overly high regard has the ability to produce poor decisions by distorting the perceived capabilities of the analyzed alternatives. In order to address this potential, an evaluation of the credibility of the risk analysis is warranted prior to deliberating the actual results.

NASA-STD-7009, Standard for Models and Simulations [31], provides the decision maker with an assessment of the modeling and simulation (M&S) results against key factors that:

- Contribute to a decision-maker's assessment of credibility and

- Are sensibly assessed on a graduated credibility assessment scale (CAS).

Table 3 (which reproduces NASA-STD-7009 Table 1) presents a high-level summary of the evaluation criteria. These are explained in greater detail in Section B.3 of the standard. Table 3 by itself is not to be used in performing credibility assessments. Rather, the detailed level definitions in the standard are to be used.

Table 3. Key Aspects of Credibility Assessment Levels
(Factors with a Technical Review subfactor are underlined)

Level	Verification	Validation	Input Pedigree	Results Uncertainty	Results Robustness	Use History	M&S Management	People Qualifications
4	Numerical errors small for all important features.	Results agree with real-world data.	Input data agree with real-world data.	Non-deterministic & numerical analysis.	Sensitivity known for most parameters; key sensitivities identified.	De facto standard.	Continual process improvement.	Extensive experience in and use of recommended practices for this particular M&S.
3	Formal numerical error estimation.	Results agree with experimental data for problems of interest.	Input data agree with experimental data for problems of interest.	Non-deterministic analysis.	Sensitivity known for many parameters.	Previous predictions were later validated by mission data.	Predictable process.	Advanced degree or extensive M&S experience, and recommended practice knowledge.
2	Unit and regression testing of key features.	Results agree with experimental data or other M&S on unit problems.	Input data traceable to formal documentation.	Deterministic analysis or expert opinion.	Sensitivity known for a few parameters.	Used before for critical decisions.	Established process.	Formal M&S training and experience, and recommended practice training.
1	Conceptual and mathematical models verified.	Conceptual and mathematical models agree with simple referents.	Input data traceable to informal documentation.	Qualitative estimates.	Qualitative estimates.	Passes simple tests.	Managed process.	Engineering or science degree.
0	Insufficient evidence.	Insufficient evidence.	Insufficient evidence.	Insufficient evidence.	Insufficient evidence.	Insufficient evidence.	Insufficient evidence.	Insufficient evidence.
	M&S Development			M&S Operations			Supporting Evidence	

To assist in the application of the evaluation criteria dictated in NASA-STD-7009, Figure 27 presents a matrix indicating the "level" of analysis of each of the estimation methods.

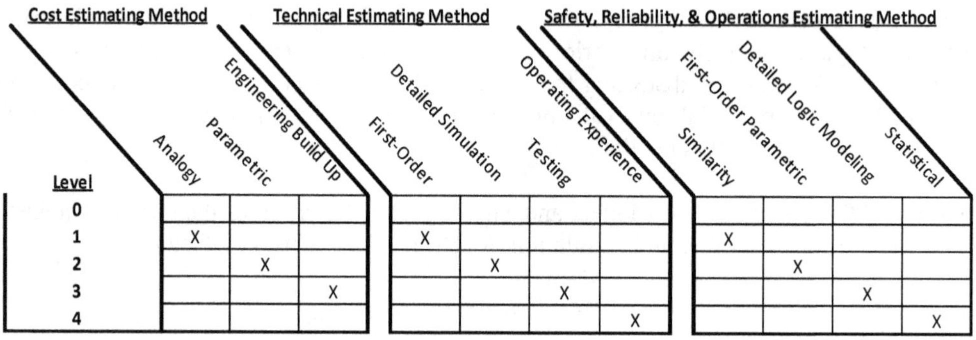

Level	Cost Estimating Method			Technical Estimating Method			Safety, Reliability, & Operations Estimating Method				
	Analogy	Parametric	Engineering Build Up	First-Order	Detailed Simulation	Testing	Operating Experience	Similarity	First-Order Parametric	Detailed Logic Modeling	Statistical
0											
1	X			X			X				
2		X			X			X			
3			X			X			X		
4										X	X

Figure 27. Analysis Level Matrix

3.2.2.8 The Technical Basis for Deliberation

The TBfD (see Appendix D) specifies the minimum information needed to risk-inform the selection of a decision alternative. The content of the TBfD is driven by the question, "What information do the deliberators and decision-makers need in order for their decision process to be fully risk-informed?"

Graphical tools are recommended, in addition to tabular data, as a means of communicating risk results. At this point in the process, the imposed constraints are the only reference points with respect to which shortfalls can be determined, so they are the only things "at risk" so far. Figure 28 presents a notional color-coded chart of imposed constraint risk. In the figure, Alternative 7 is relatively low risk for every listed performance measure (i.e., those with imposed constraints on the allowable values), as well as for all constrained performance measures collectively (the "Total" column). Alternatives 12 and 3 have a mix of performance measure risks, some of which are high, resulting in a high risk of failing to meet one or more imposed constraints.

To assist the deliberators and decision-maker in focusing on the most promising alternatives, with an awareness of the relative risks to imposed constraints, the imposed constraints risk matrix has been:

- Sorted by total risk, with the least risky alternatives at the top; and

- Colored on a relative basis from low risk (the blue-violet end of the spectrum) to high risk (the orange-red end of the spectrum).

When presenting the performance measure pdfs themselves, "band-aid" charts can be used, which show the mean, 5th percentile, and 95th percentile values (and often the median as well). Figure 29 shows a notional example of a band-aid chart. Unlike the imposed constraints matrix, which includes only those performance measures that have imposed constraints, band-aid charts can be made for every performance measure in the risk analysis, thereby giving a complete picture of the analyzed performance of each alternative.

Alternative	Imposed Constraints					Total
	PM$_1$	PM$_2$	PM$_3$	...	PM$_n$	
	Constraint (> C$_1$)	Constraint (< C$_2$)	Constraint (> C$_3$)		Constraint (< C$_n$)	
7	0.7%*	0.09%	0.04%		0.1%	0.9%
1	2%	0.8%	0.2%		0.01%	4%
12	0.4%	5%	20%		0.1%	27%
...				...		
3	15%	0.9%	8%		0.01%	56%

*The probability of not meeting the imposed constraint

Figure 28. Notional Imposed Constraints Risk Matrix

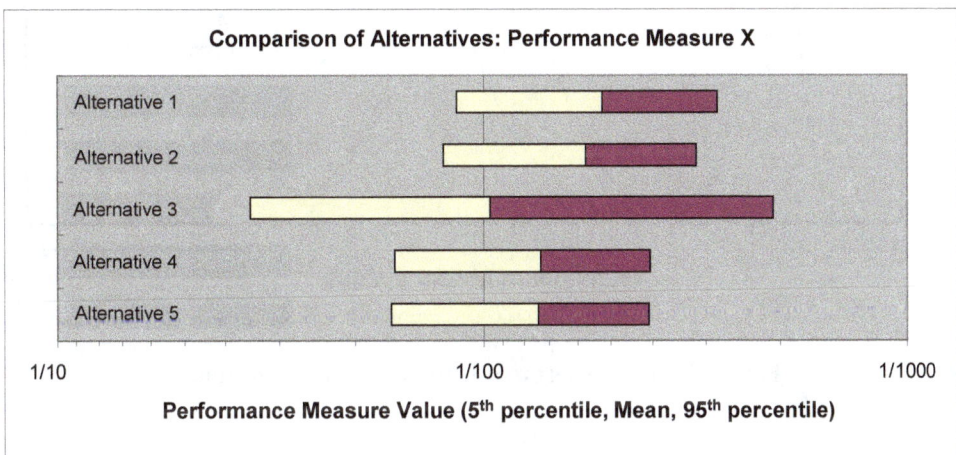

Comparison of Alternatives: Performance Measure X

Alternative 1
Alternative 2
Alternative 3
Alternative 4
Alternative 5

1/10   1/100   1/1000

Performance Measure Value (5th percentile, Mean, 95th percentile)

Figure 29. Notional Band Aid Chart for Performance Measure X

When using charts such as the band-aid chart of Figure 29, it is important to know the degree of correlation among the different alternatives. For example, in the figure, the pdfs of Alternative 1 and Alternative 2 overlap to an extent that it may seem that the chances of either one having the higher performance are about the same. Indeed, this is true if the pdfs are independent. However, if they are correlated, then it might not be the case. For example, suppose the alternatives are identical except for some small design difference that slightly increases the value of Performance Measure X for Alternative 2. Then, although the performance of both alternatives is uncertain, the performance difference between them is known and constant.

A direct representation of the difference between design alternatives, including the associated uncertainty, can supplement the information provided by band-aid charts, allowing for a better ability to make comparisons under uncertainty. A possible representation is shown in Figure 30 [32]. The figure shows performance measure pdfs for two alternatives whose performance measure values are correlated. A third, dotted, curve shows the pdf of the performance difference between the two alternatives. This curve indicates that despite the significant overlap between the two performance measure pdfs, Alternative 2 is unequivocally superior to Alternative 1, at least for the performance measure shown.

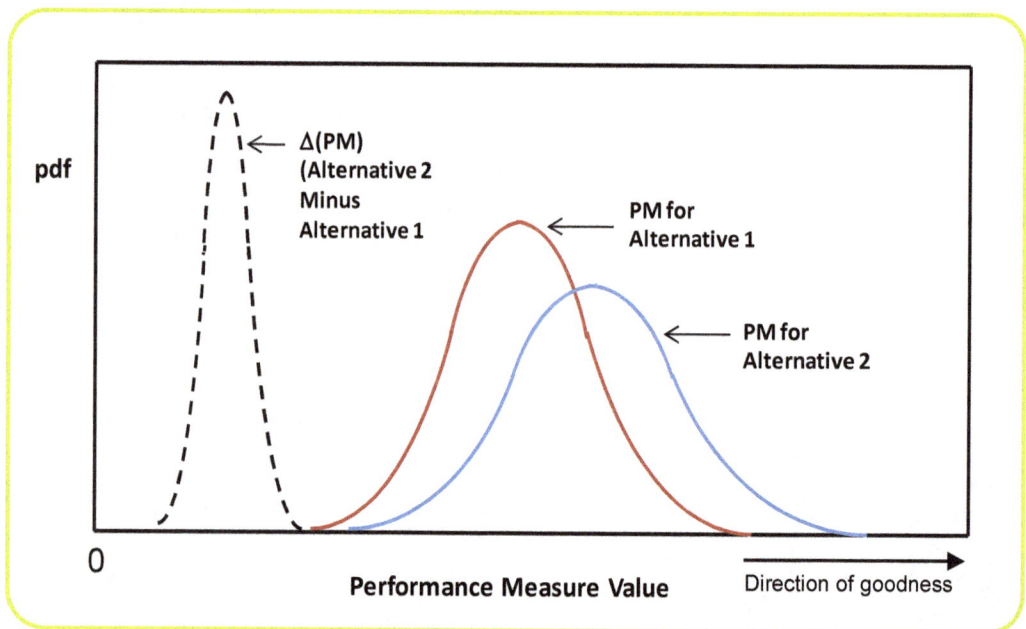

Figure 30. Comparison of Uncertainty Distributions

Planetary Science Mission Example: Conduct the Risk Analysis

Each of the modeled performance measures is quantified, using a Monte Carlo shell to sample the uncertain performance parameter pdfs and propagate them through the analysis framework, producing the performance measure results shown below. Each chart presents the assessed performance of the alternatives for a single performance measure, as well as the applicable imposed constraint, which defines the level of performance needed in order to fully meet the top-level objectives.

The risk analysis of alternatives produced the following pdfs:

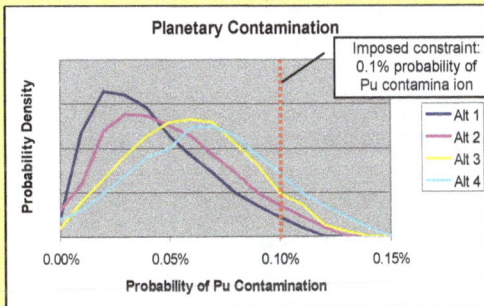

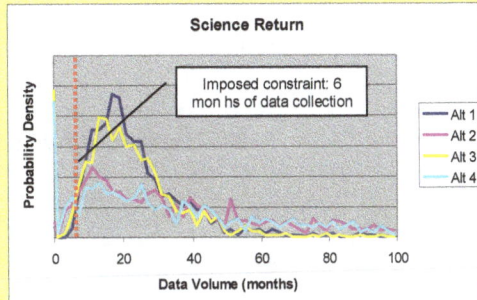

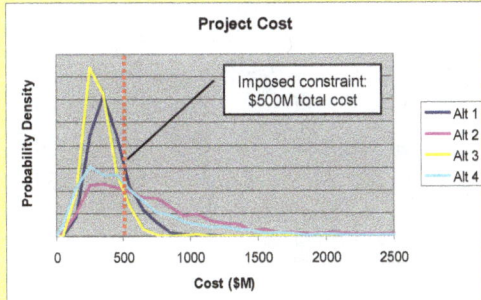

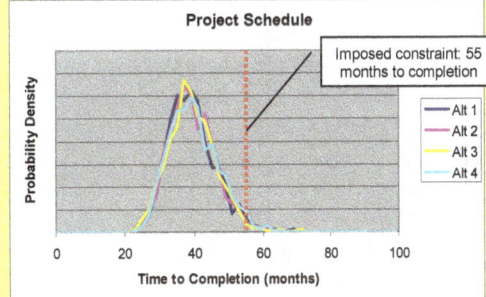

Performance Measure Results for the Planetary Science Mission Example

There is substantial overlap among the pdfs, particularly for *time to completion*. In practice, consideration is given to performing additional analysis to resolve such overlap in cases where doing so is expected to illuminate the decision. However, additional analysis might not help to distinguish among alternatives, especially when the underlying uncertainties are common to them. This is considered to be the case for the notional analysis of the Planetary Science Mission Example.

3.3   Part 3 – Risk-Informed Alternative Selection

The risk-informed alternative selection process within RIDM provides a method for integrating risk information into a deliberative process for decision making, relying on the judgment of the decision-makers to make a risk-informed decision. The decision-maker does not necessarily base his or her selection of a decision alternative solely on the results of the risk analysis. Rather, the risk analysis is just one input to the process, in recognition of the fact that it may not model everything of importance to the stakeholders. Deliberation employs critical thinking skills to the collective consideration of risk information, along with other issues of import to the stakeholders and the decision-maker, to support decision making.

Figure 31 illustrates Part 3 of the RIDM process, Risk-Informed Alternative Selection.

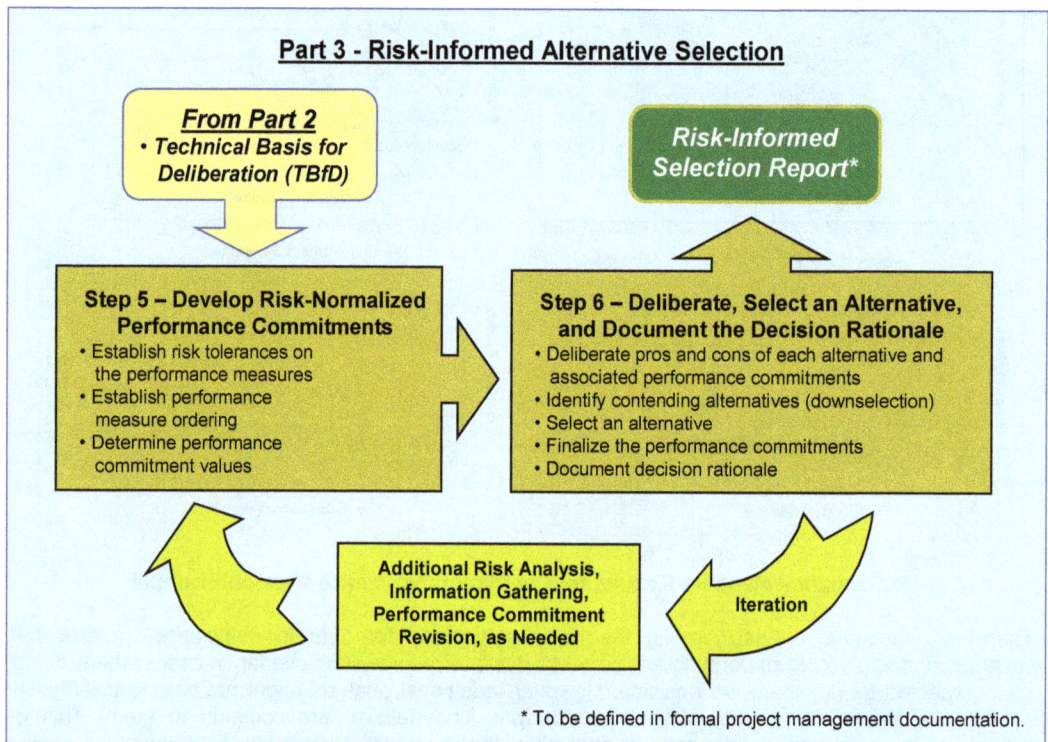

Figure 31. RIDM Process Part 3, Risk-Informed Alternative Selection

In Step 5, performance commitments are developed, representing consistent levels of risk tolerance across alternatives. In Step 6, relevant stakeholders, risk analysts, and decision-makers deliberate the relative merits and drawbacks of each alternative, given the information in the TBfD. This step is iterative, and may involve additional risk analysis or other information gathering as the participants strive to fully assess the alternatives and identify those that they consider to be reasonable contenders, worthy of serious consideration by the decision-maker. The decision-maker, or his/her proxy, may also be involved at this stage to help cull the number of alternatives (a.k.a. downselecting). Once a set of contending alternatives has been identified, the decision-maker integrates the issues raised during deliberation into a rationale for the selection of an alternative, and finalizes the performance commitments. The decision rationale is then documented in accordance with any existing project management directives, in a RISR. For pedagogical purposes, the process is laid out as if alternative selection involves a single decision that is made once deliberations are complete. However, as discussed in the Section 3.2.2.4 guidance on sequential analysis and downselection, decisions are often made in stages and in a number of forums that may involve a variety of proxy decision-makers.

Additionally, this handbook refers to the participants in deliberation as deliberators. This is also for pedagogical purposes. As illustrated in Figure 5, deliberators may be drawn from any of the sets of stakeholders, risk analysts, SMEs, and decision-makers.

3.3.1 Step 5 – Develop Risk-Normalized Performance Commitments

In order to generalize the consistent application of risk tolerance to the performance expectations of each decision alternative, this handbook introduces the concept of performance commitments. A performance commitment is a performance measure value set at a particular percentile of the performance measure's pdf, so as to anchor the decision-maker's perspective to that performance measure value as if it would be his/her commitment, were he/she to select that alternative. For a given performance measure, the performance commitment is set at the same percentile for all decision alternatives, so that the probability of failing to meet the different alternative commitment values is the same across alternatives.

Performance commitments support a *risk-normalized* comparison of decision alternatives, in that a uniform level of risk tolerance is established prior to deliberating the merits and drawbacks of the various alternatives. Put another way, risk-normalized performance commitments show what each alternative is capable of with an equal likelihood of achieving that capability, given the state of knowledge at the time.

The inputs to performance commitment development are:

- The performance measure pdfs for each decision alternative;

- An ordering of the performance measures; and

- A risk tolerance for each performance measure, expressed as a percentile value.

For each alternative, performance commitments are established by sequentially determining, based on the performance measure ordering, the value that corresponds to the stated risk tolerance, conditional on meeting previously-defined performance commitments. This value becomes the performance commitment for the current performance measure, and the process is repeated until all performance commitments have been established for all performance measures. Figure 32 illustrates the process.

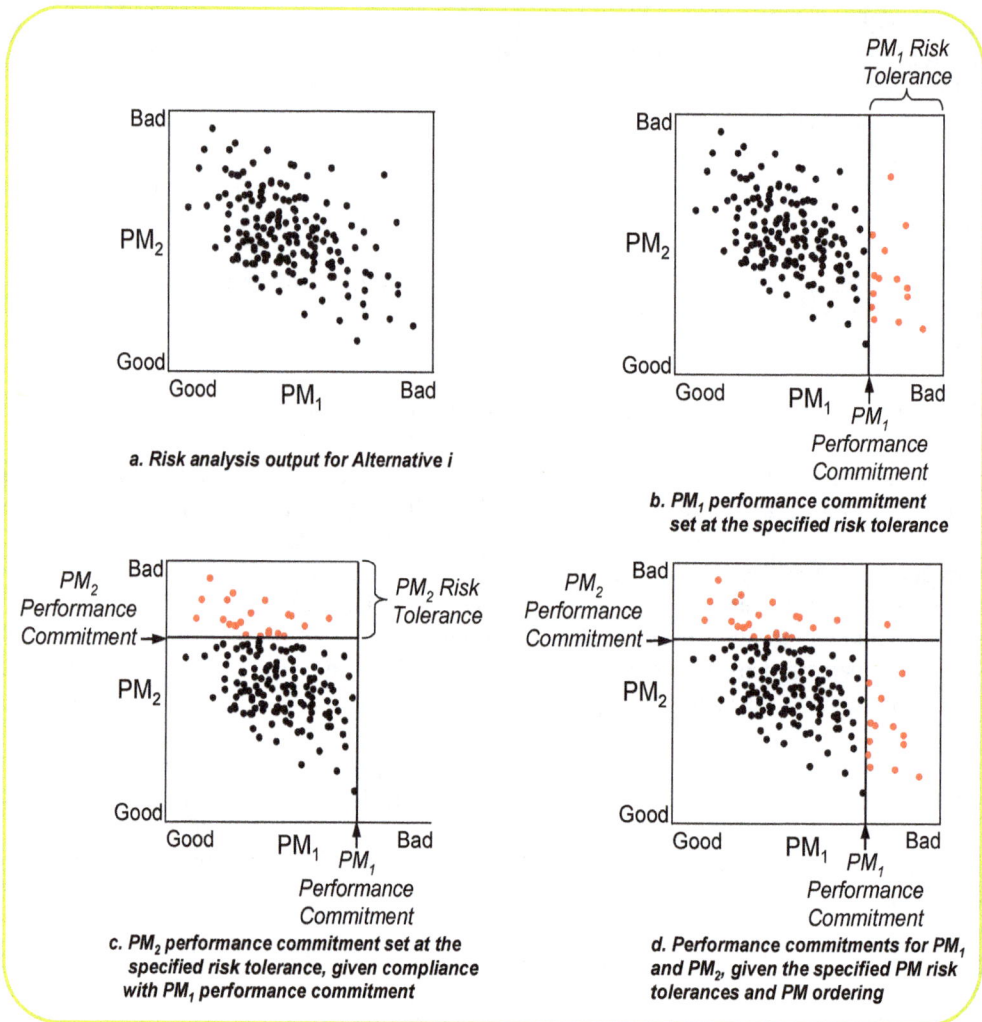

Figure 32. Establishing Performance Commitments

In Figure 32 there are only two performance measures, PM_1 and PM_2. If, for example, PM_1 is P(LOC) and PM_2 is cost, then, the risk analysis results can be shown as a scatter plot on the P(LOC)-cost plane (see Figure 32a), where each point represents the output from a single iteration of the Monte Carlo shell. If the ordering of the performance measures is P(LOC) first and cost second, P(LOC) would be the first performance measure to have a performance commitment established for it (see Figure 32b). This is done by determining the value of P(LOC) whose probability of exceedance equals the defined risk tolerance. That value becomes the P(LOC) performance commitment.[7] The process is repeated for cost, *conditional on the P(LOC) performance commitment being met*. Thus, the points on the scatter plot that exceed the P(LOC)

---

[7] If the "direction of goodness" of the performance measure were reversed, the performance commitment would be at the value whose probability of exceedance equals one minus the risk tolerance.

performance commitment have been removed from consideration and the cost performance commitment is established solely on the basis of the remaining data (see Figure 32c). The result is a set of performance commitments for the P(LOC) and cost performance measures that reflects the risk tolerances of the deliberators and decision-maker (see Figure 32d). This procedure can be extended to any number of performance measures.

In general, different decision alternatives will have different performance commitments. But the probability of meeting each performance commitment will be the same (namely, one minus the risk tolerance of that performance measure), given that prior performance commitments in the performance measure ordering have been met:

$$\text{P(Performance Commitment i is met)} = 1 - \text{PM}_i \text{ Risk Tolerance}$$

$$= 1 - \text{P(Performance Commitment i is unmet} \mid \text{Performance Commitments j} < \text{i are met)}$$

Moreover, the probability of meeting all performance commitments is identical for all alternatives, and is calculated as:

$$\text{P(All Performance Commitments Met)} = \prod_{i=1}^{\#PMs} (1 - \text{PM}_i \text{ Risk Tolerance})$$

3.3.1.1 Establishing Risk Tolerances on the Performance Measures

The RIDM process calls for the specification of a risk tolerance for each performance measure, along with a performance measure ordering, as the basis for performance commitment development. These risk tolerance values have the following properties:

- The risk tolerance for a given performance measure is the same across all alternatives, and

- Risk tolerance may vary across performance measures, in accordance with the stakeholders' and decision-maker's attitudes towards risk for each performance measure.

Risk tolerances, and their associated performance commitments, play multipurpose roles within the RIDM process:

- Uniform risk tolerance across alternatives normalizes project/program risk, enabling deliberations to take place that focus on performance capabilities on a risk-normalized basis.

- The risk tolerances that are established during the RIDM process indicate the levels of acceptable initial risk that the CRM process commits to managing during implementation. (Note: The *actual* initial risk is not established until performance requirements are agreed upon as part of the overall systems engineering process, and not explicitly addressed until the CRM process is initialized. More information on CRM initialization can be found in Section 4.)

- Performance commitments based on risk tolerance enable point value comparison of alternatives in a way that is appropriate to a situation that involves thresholds (e.g., imposed constraints). By comparing a performance commitment to a threshold, it is immediately clear whether or not the risk of crossing the threshold is within the established risk tolerance. In contrast, if a value such as the distribution mean were used to define performance commitments, the risk with respect to a given threshold would not be apparent.

Issues to consider when establishing risk tolerances include:

- *Relationship to imposed constraints* – In general, deliberators have a low tolerance for noncompliance with imposed constraints. Imposed constraints are akin to the success criteria for top-level objectives; if imposed constraints are not met, then objectives are not met and the endeavor fails. By establishing a correspondingly low risk tolerance on performance measures that have imposed constraints, stakeholders and decision-makers have assurance that if an alternative's performance commitments exceed the associated imposed constraints, there is a high likelihood of program/project success.

- *High-priority objectives* – It is expected that deliberators will also have a low risk tolerance for objectives that have high priority, but for which imposed constraints have not been set. The lack of an imposed constraint on a performance measure does not necessarily mean that the objective is of less importance; it may just mean that there is no well defined threshold that defines success. This could be the case when dealing with quantities of data, sample return mass capabilities, or operational lifetimes. It is generally the case for life safety, for which it is difficult to establish a constraint *a priori*, but which is nevertheless always among NASA's top priorities.

- *Low-priority objectives and/or "stretch goals"* – Some decision situations might involve objectives that are not crucial to program/project success, but which provide an opportunity to take risks in an effort to achieve high performance. Technology development is often in this category, at least when removed from a project's critical path. In this case, a high risk tolerance could be appropriate, resulting in performance commitments that suggest the alternatives' performance potentials rather than their established capabilities.

- *Rebaselining issues* – Requirements on some performance measures might be seen as difficult to rebaseline. For these performance measures, deliberators might establish a low risk tolerance in order to reduce the possibility of having to rebaseline.

Risk tolerance values are up to the deliberators and decision maker, and are subject to adjustment as deliberation proceeds, opinions mature, and sensitivity excursions are explored. In particular, it is recommended that sensitivity excursions be explored over a reasonable range of risk tolerances, not only for the purpose of making a decision that is robust with respect to different risk tolerances, but also in order to find an appropriate balance between program/project risk and the performance that is specified by the performance commitments.

3.3.1.2 Ordering the Performance Measures

Because of possible correlations between performance measures, performance commitments are developed sequentially. As discussed in Section 3.3.1, performance commitments are defined at the value of a performance measure that corresponds to the defined risk tolerance, conditional on meeting previously defined performance commitments. In general, performance commitments depend on the order in which they are developed.

Qualitatively, the effect that performance measure order has on performance commitment values is a follows:

- If performance measures are independent, then the order is immaterial and the performance commitments will be set at the defined risk tolerances of the performance measures' marginal pdfs.

- If performance measures are positively correlated in terms of their directions of goodness, then the performance commitments that lag in the ordering will be set at higher levels of performance than would be suggested by their marginal pdfs alone. This is because lagging performance measures will have already been conditioned on good performance with respect to leading performance measures. This, in turn, will condition the lagging performance measures on good performance, too, due to the correlation.

- If performance measures are negatively correlated in terms of their directions of goodness, then the performance commitments that lag in the ordering will be set at lower levels of performance than would be suggested by their marginal pdfs alone. Figure 32 shows this phenomenon. In Figure 32c, the PM_2 performance commitment is set at a slightly lower performance than it would have been if the data points that exceed the PM_1 performance commitment were not "conditioned out."

- The lower the risk tolerance, the lower the effect of conditioning on subsequent performance commitments. This is simply because the quantity of data that is "conditioned out" is directly proportional to risk tolerance.

These general effects of performance measure ordering on performance commitments suggest the following ordering heuristics:

- Order performance measures from low risk tolerance to high risk tolerance. This assures a minimum of difference between the risk tolerances as defined on the conditioned pdfs versus the risk tolerances as applied to the marginal pdfs.

- Order performance measures in terms of the desire for specificity of the performance measure's risk tolerances. For example, the performance commitment for the first performance measure in the ordering is precisely at its marginal pdf. As subsequent performance commitments are set, dispersion can begin to accumulate as conditioning increases.

Once the performance commitments are developed, each alternative can be compared to every other alternative in terms of their performance commitments, with the deliberators understanding that the risk of not achieving the levels of performance given by the performance commitments is the same across alternatives. Additionally, the performance commitments can be compared to any imposed constraints to determine whether or not the possibility that they will not be satisfied is within the risk tolerance of the deliberators, and ultimately, the decision maker. Figure 33 notionally illustrates a set of performance commitments for each of three competing alternatives. Note that Alternative A does not satisfy the imposed constraint on payload capability within the risk tolerance that has been established for that performance measure.

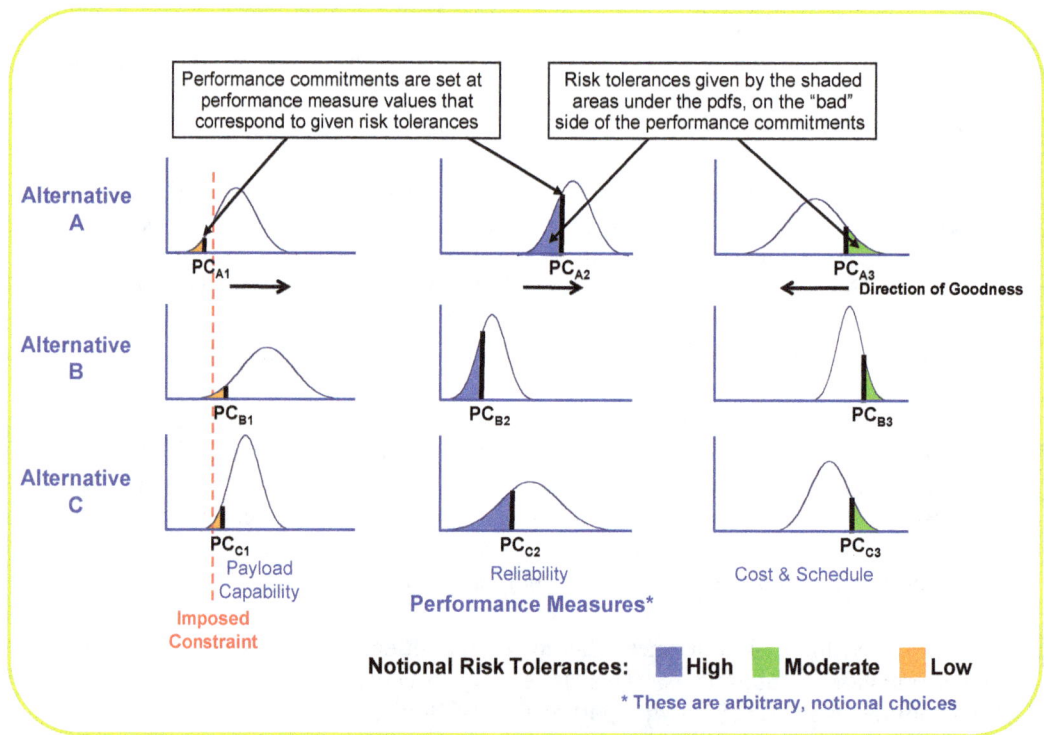

Figure 33. Performance Commitments and Risk Tolerances for Three Alternatives

3.3.2 Step 6 – Deliberate, Select an Alternative, and Document the Decision Rationale

The RIDM process invests the decision-maker with the authority and responsibility for critical decisions. While ultimate responsibility for alternative selection rests with the decision-maker, alternative evaluation can be performed within a number of deliberation forums which may be held before the final selection is made. As partial decisions or "down-selects" may be made at any one of these deliberation forums, they are routinely structured around a team organizational structure identified by the decision-maker. It is important to have a team with broad based expertise to perform sufficient analysis to support a recommendation or decision. At the top of the structure may be the decision-maker or a deliberation lead appointed by the decision-maker. If a deliberation lead is appointed this individual should be an experienced manager, preferably one with an analytical background.

3.3.2.1 Convening a Deliberation Forum

Deliberation forums address the major aspects of the decision. The use of these forums helps ensure that a responsible person leads each important area of analysis. The focus of these forums will vary with the type of study.

Depending on circumstances, forums can be split (e.g., into separate Safety and Technical), or functions can be combined (e.g., Cost and Schedule), or entirely new forums can be created (e.g., test, requirements or stakeholder). The final choice of forum structure belongs to the decision-maker. At a minimum, the forums should mirror the major aspects of the study. Thus the creation of forums offers an important early opportunity to contemplate the efforts processes and goals. Every forum must have enough members to achieve a "critical mass" of knowledge, interest and motivation. Typically, a small group with critical mass is more productive than a larger group with critical mass. This suggests starting with a small forum and adding members as necessary.

Members of a deliberation forum should ideally be selected based on their qualifications. Consideration should be given to those with relevant experience, knowledge, and interest in the subject matter. These individuals are frequently referred to as SMEs. In some cases they have an organizational charter to support the process and in other cases they participate because they are heavily invested in the outcome of the deliberation. When the most qualified are not available, the next most qualified should be sought.

People with diverse viewpoints on controversial issues should also be enlisted to participate in deliberations. They should represent the diversity of stakeholder interests. Partisans, by their nature, will defend their ideas and detect flaws in the ideas of their competition. This allows issues to be raised and resolved early that might otherwise lie in wait. A formal tracking system should be employed throughout the process to track items to closure.

Additional information on deliberative processes can be found in [33].

Planetary Science Mission Example: Develop Risk-Normalized Performance Commitments

Risk-normalized performance commitments were developed for each of the four analyzed alternatives, for the performance measures of *time to completion*, *project cost*, *data volume*, and *planetary contamination*.

The table below shows the risk tolerance given to each. Because of the importance of meeting the 55 month launch window, a low risk tolerance of 3% is given to *time to completion*. Then, given the 3% *time to completion* risk tolerance, a 27% risk tolerance was given to *project cost* based on the NASA policy of budgeting cost and schedule at a joint confidence level (JCL) of 70% [7]. The 10% *data volume* risk tolerance is reasonably low, but reflects the belief that minor shortfalls will not significantly erode the success of the mission. The *planetary contamination* risk tolerance is moderately low, reflecting the importance of avoiding radiological releases into the planetary environment. These risk tolerance values are discussed with the decision maker to ensure that they are consistent with his/her views.

The table also shows the ordering of the performance measures that was used to develop performance commitments. *Time to completion* was chosen first due to its critical importance. *Project cost* was chosen next, due to its importance in an environment of scarce resources, and also because of its linkage to *time to completion* via the NASA JCL policy. Data volume was chosen third, due to its prominence among the technical objectives.

**Performance Measure Risk Tolerance for
Planetary Science Mission Example**

Performance Measure	Risk Tolerance	Performance Measure Ordering
Time to Completion	3%	1
Project Cost	27%	2
Data Volume	10%	3
Planetary Contamination	15%	4

The performance commitment chart (on the next page) shows the levels of performance that are achievable at the stated risk tolerances. One thing that is immediately evident is that Alternative 4 does not meet the 6 month *data volume* imposed constraint. However, the deliberation team recognizes that a different risk tolerance might produce a *data volume* performance commitment that is in line with the imposed constraint, so they are reluctant to simply discard Alternative 4 out of hand. Instead, they determine the risk that would have to be accepted in order to produce a *data volume* performance commitment of at least 6 months. This turns out to be 12%, which the team considers to be within the range of reasonable tolerances (indeed, it is not significantly different from 10%). In the interest of illuminating the situation to the decision maker, the team includes both sets of results in the chart.

Planetary Science Mission Example: Develop Risk-Normalized Performance Commitments (continued)

Planetary Science Mission Example - Performance Commitment Chart

Risk Tolerance	Performance Commitments			
	Time to Completion (months)	Project Cost ($M)	Data Volume (months)	Planetary Contamination (probability)
Alternative	3%	27%	10% (12%)	15%
1. Propulsive Insertion, Low-Fidelity Science Package	54	576	11 (11)	0.07%
2. Propulsive Insertion, High-Fidelity Science Package	53	881	9.9 (11)	0.08%
3. Aerocapture, Low-Fidelity Science Package	55	413	6.8 (7.8)	0.10%
4. Aerocapture, High-Fidelity Science Package	54	688	4.9 (6.0)	0.11%

3.3.2.2 Identify Contending Alternatives

After the performance commitments have been generated, they are used to pare down the set of decision alternatives to those that are considered to be legitimate contenders for selection by the decision-maker. This part of the process is a continuation of the pruning activity begun in Section 3.1.2. At this point, however, the deliberators have the benefit of the TBfD and the performance commitments, as well as the subjective, values-based input of the deliberators themselves. Rationales for elimination of non-contending alternatives include:

- **Infeasibility** – Performance commitments are exceeded by the imposed constraints. In this case, imposed constraints cannot be met within the risk tolerance of the decision-maker.

- **Dominance** – Other alternatives exist that have superior performance commitments on every performance measure, and substantially superior performance on some.[8] In this case, an eliminated alternative may be feasible, but nonetheless is categorically inferior to one or more other alternatives.

---

[8] When eliminating alternatives on the basis of dominance, it is prudent to allow some flexibility for uncertainty considerations beyond those captured by the performance commitments alone (discussed in the next subsection). Minor performance commitment shortfalls relative to other alternatives do not provide a strong rationale for elimination, absent a more detailed examination of performance uncertainty.

- **Inferior Performance in Key Areas** – In general, in any decision involving multiple objectives, some objectives will be of greater importance to deliberators than others. Typically important objectives include crew safety, mission success, payload capability, and data volume/quality. Alternatives that are markedly inferior in terms of their performance commitments in key areas can be eliminated on that basis, in recognition of stakeholder and decision-maker values.

Section 3.2.2.4 discusses sequential analysis and downselection, in which non-contending alternatives are identified and eliminated in parallel with risk analysis, thereby reducing the analysis burden imposed by the decision-making process. Sequential analysis and downselection represents a graded approach to the identification of contending alternatives, and is another example of the iterative and collaborative nature of the RIDM process.

3.3.2.3 Additional Uncertainty Considerations

The guidance above for identifying contending alternatives is primarily focused on comparisons of performance commitments. This facilitates comparisons between alternatives (and against imposed constraints), and the elimination of non-contenders from further consideration. However, performance commitments do not capture all potentially relevant aspects of performance, since they indicate the performance at only a single percentile of each performance measure pdf. Therefore, alternatives identified as contenders on the basis of their performance commitments are further evaluated on the basis of additional uncertainty considerations relating to their performance at other percentiles of their performance measure pdfs. In particular, performance uncertainty may give rise to alternatives with the following characteristics:

- **They offer superior expected performance** – In many decision contexts (specifically, those in which the decision-maker is *risk neutral*[9]), the decision-maker's preference for an alternative with uncertain performance is equivalent to his or her preference for an alternative that performs at the mean value of the performance measure pdf. When this is the case, expected performance is valuable input to decision making, as it reduces the comparison of performance among alternatives to a comparison of point values.

  However, in the presence of performance thresholds, over-reliance on expected performance in decision making has the potential to:

  o Introduce potentially significant probabilities of falling short of imposed constraints, thereby putting objectives at risk, even when the mean value meets the imposed constraints

  o Contribute to the development of derived requirements that have a significant probability of not being achievable

---

[9] A risk-neutral decision maker is indifferent towards a decision between an alternative with a definite performance of X, versus an alternative having an uncertain performance whose mean value is X. In other words, a risk-neutral decision maker is neither disproportionally attracted to the possibility of exceptionally high performance (*risk seeking*) nor disproportionally averse to the possibility of exceptionally poor performance (*risk averse*).

Since direction-setting, requirements-producing decisions at NASA typically involve performance thresholds, expected performance should be considered in conjunction with performance commitments, to assure that the decision is properly risk informed.

- **They offer the potential for exceptionally high performance** – For a given performance measure pdf, the percentile value at the decision-maker's risk tolerance may be unexceptional relative to other contending alternatives. However, at higher risk tolerances, its performance may exceed that of other alternatives, to the extent that it becomes attractive relative to them. This may be the case even in the presence of inferior performance commitments on the same, or different, performance measures.

An example of this is shown notionally in Figure 34. In this figure, Alternative 2's performance commitment is at a worse level of performance than Alternative 1's; however, Alternative 2 offers a possibility of performance that is beyond the potential of Alternative 1. In this case, stakeholders and decision-makers have several choices. They can:

  o Choose Alternative 1 on the basis of superior performance at their risk tolerance;

  o Choose Alternative 2 on the basis that its performance at their risk tolerance, though not the best, is acceptable, and that it also has the potential for far superior performance; or

  o Set their risk tolerance such that the performance commitment for both alternatives is the same thus making this performance measure a non-discriminator between the two options.

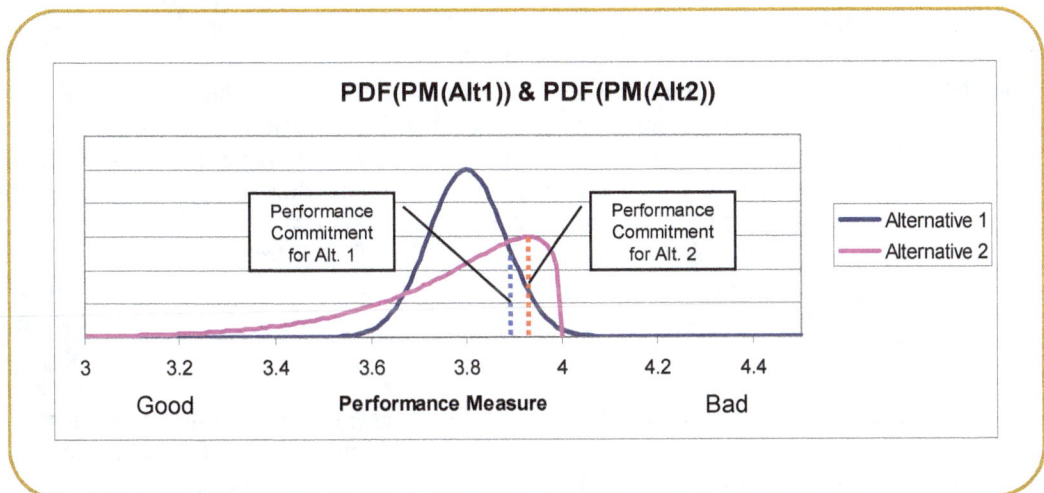

Figure 34. An Example Uncertainty Consideration: The Potential for High Performance

In the second case, the decision-maker is accepting a higher program/project risk, which will lead to the development of more challenging requirements and increased CRM burden regardless of which alternative is selected.

- **They present a risk of exceptionally poor performance** – This situation is the reverse of the situation above. In this case, even though the likelihood of not meeting the performance commitment is within the decision-makers' risk tolerance, the consequences may be severe, rendering such an alternative potentially unattractive.

Another uncertainty consideration, which is addressed below in the discussion of the iterative nature of deliberation, is whether or not a performance measure's uncertainty can be effectively reduced, and whether or not the reduction would make a difference to the decision. This issue is mentioned here because wide pdfs can lead to poor performance commitments relative to other alternatives, and it would be unfortunate to discard an alternative on this basis if additional analysis could be done to reduce uncertainty. Note that if two attractive alternatives present themselves and time and resources are available, it may be advantageous to proceed with, at least, partial prototyping (that is, prototyping of some of the critical components) of both to provide the necessary data for reducing key performance measure uncertainties such that a robust decision can be made.

3.3.2.4 Other Considerations

Depending on the decision situation and proposed alternatives, a variety of other risk-based, as well as non-risk-based, considerations may also be relevant. These include:

- **Sensitivity of the performance commitments to variations in risk tolerance** – Performance commitments are directly related to risk tolerance. Therefore, it is prudent for the deliberators to explore the effects of variations in the specified risk tolerances, to assure that the decision is robust to variations within a reasonable range of tolerances.

- **Risk disposition and handling considerations** – The risks that exist relative to performance commitments are ultimately caused by undesirable scenarios that are identified and analyzed in the risk analysis. Because of the scope of risk analysis for RIDM (i.e., the necessity to analyze a broad range of alternatives), risk retirement strategies may not be fully developed in the analysis. Deliberators' expertise is therefore brought to bear on the relative risk-retirement burdens that different alternatives present. For example, deliberators might feel more secure accepting a technology development risk that they feel they can influence, rather than a materials availability risk they are powerless to control.

- **Institutional considerations** – Different alternatives may have different impacts on various NASA and non-NASA organizations and institutions. For example, one alternative might serve to maintain a particular in-house expertise, while another alternative might help maintain a regional economy. These broad-ranging issues are not necessarily captured in the performance measures, and yet they are of import to one or more stakeholders. The deliberation forum is the appropriate venue for raising such issues for formal consideration as part of the RIDM process.

3.3.2.5 Deliberation Is Iterative

As illustrated in Figure 31, deliberation is an iterative process that focuses in on a set of contending alternatives for consideration by the decision-maker. Iteration during deliberation has both qualitative and quantitative aspects:

- **Qualitative** – A deliberator may have a particular issue or concern that he or she wishes to reach closure on. This might require several rounds of deliberation as, for example, various subject matter experts are called in to provide expertise for resolution.

- **Quantitative** – One or more performance measures might be uncertain enough to significantly overlap, thereby inhibiting the ability to make a robust decision. Moreover, large uncertainties will, in general, produce poor performance commitments, particularly when risk tolerance is low. Therefore, before a set of contending alternatives can be chosen, it is important that the deliberators are satisfied that particular uncertainties have been reduced to a level that is as low as reasonably achievable given the scope of the effort. It is expected that the risk analysis will be iterated, under the direction of the deliberators, to address their needs.

3.3.2.6 Communicating the Contending Alternatives to the Decision Maker

There comes a time in RIDM when the remaining alternatives all have positive attributes that make them attractive in some way and that make them all contenders. The next step is to find a way to clearly state for the decision-maker the advantages and disadvantages of each remaining alternative, especially how the alternatives address imposed constraints and satisfy stakeholder expectations. It is important that the process utilized by the deliberators affords him or her with ample opportunity to interact with the deliberators in order to fully understand the issues. This is particularly true if the decision-maker has delegated deliberation and downselection to a proxy. The information and interaction should present a clear, unbiased picture of the analysis results, findings, and recommendations. The more straightforward and clear the presentation, the easier it becomes to understand the differences among the alternatives.

Some of the same communication tools used in the TBfD can be used here as well, applied to the contending alternatives forwarded for the decision-maker's consideration. The imposed constraints risk matrix (Figure 28) summarizes what is among the most critical risk information. Additionally, information produced during deliberation should be summarized and forwarded to the decision-maker. This includes:

- **Risk tolerances and performance commitments** – The deliberators establish risk tolerances on the performance measures, for the purpose of generating performance commitments that can serve as the primary basis for comparison of alternatives. These tolerances and the resulting performance commitments are key pieces of information for the decision-maker. They strongly influence requirements development and the corresponding program/project risk that is to be accepted going forward. A notional performance commitment chart is shown in Figure 35.

Risk Tolerance	Performance Commitments					Performance Legend
	PM_1	PM_2	PM_3		PM_n	
Alternative	10%	1%	30%	...	5%	Lower
1	100	0.007	0.01		0.4	Medium
2	500	0.0002	0.15		5	Higher
3	350	0.004	0.08		20	
...				...		
m	400	0.0001	0.25		0.1	

Figure 35. Notional Performance Commitment Chart

- **Pros and cons of each contending alternative** – An itemized table of the pros and cons of each alternative is also recommended for the contending alternatives. This format has a long history of use, and is capable of expressing qualitative issues. It enables conflicting opinions to be documented and communicated to the decision-maker, so that he or she is aware of contentious issues and/or competing objectives among stakeholders.

- **Risk lists** – Each alternative will have different contributors to its performance commitment risks. Correspondingly, each contending alternative will have a risk list written for it that identifies the major scenarios that contribute to risk. Each scenario has the potential to impact multiple performance measures over multiple mission execution domains.

Figure 36 presents a notional example of a RIDM risk list. Each row of Figure 36 represents a "risk," as the term is used in the CRM process. Each risk is articulated in a risk statement, which identifies an existing *condition* (e.g., "A" for Risk #1) that indicates a possibility of some future *consequence* ("B" for Risk #1) that contributes to one or more performance commitments not being met. The magnitude of the contribution is indicated in stoplight format (red/yellow/green) on a performance commitment basis, as well as on a holistic basis. The basis for determining the magnitude depends on the form of the risk assessment and the criteria established in the risk management plan (RMP), if one exists. For example, analyses that use detailed logic modeling might express risk contributions in terms of importance measures such as the Fussell-Vesely or Risk

Reduction Worth (RRW) importance measures [24]. Less detailed analyses might use more qualitative criteria. Whatever method is used, consistency between the RIDM and CRM processes in this respect aids in the initialization of CRM for the selected alternative.

Alternative X – RIDM Risk Analysis – Risk List								
Risk #	Risk Statement	Performance Commitments					Total	Risk Legend
		PM_1	PM_2	PM_3	...	PM_n		
1	Given A there is a possibility of B							High
2	Given C there is a possibility of D							Medium
3	Given E there is a possibility of F							Low
...				...				N/A
m	Given Y there is a possibility of Z							

Figure 36. Notional Risk List for Alternative X

Regardless of how well the risk information is summarized or condensed into charts or matrices, the decision-maker should also always be presented with the raw risk results, namely the performance measure pdfs, upon request. Only by having these fundamental analysis results can the decision-maker bring his or her full judgment to bear on the selection of an alternative. Band-aid charts, as shown in Figure 29, are appropriate communication tools for communicating this information to the decision-maker.

3.3.2.7 Alternative Selection Is Iterative

Just as risk analysis and deliberation iterate until the deliberators are satisfied that their issues and concerns have been satisfactorily addressed, alternative selection also iterates until the decision-maker is satisfied that the information at his or her disposal is sufficient for making a risk-informed decision. This is especially true in situations where the decision-maker has delegated much of the activity to others, and is exposed to the issues mainly through summary briefings of analyses and deliberations conducted beforehand. Iteration might consist of additional focused analyses, additional subject matter expert input, consideration of alternate risk tolerances (and associated performance commitments) for some performance measures, etc.

3.3.2.8 Selecting a Decision Alternative

Once the decision-maker has been presented with enough information for risk-informed decision making, he or she is ready to select a decision alternative for implementation. The decision itself consists of two main ingredients: the selection of the decision alternative and finalization of the performance commitments.

- **Selecting a decision alternative** – The RIDM process is concerned with assuring that decisions are risk-informed, and does not specify a particular process for selecting the decision alternative itself. Decision-makers are empowered to use their own methods for decision making. These may be qualitative or quantitative; they may be structured or unstructured; and they may involve solitary reflection or the use of advisory panels. Regardless of the method used for making the decision, the decision-maker formulates and documents the decision rationale in light of the risk analysis.

- **Finalizing the performance commitments** – In the requirements-based environment of the NASA program/project life cycle, decisions are essentially defined by the requirements they produce. Performance commitments capture the performance characteristics that the decision-maker expects from the implemented alternative, and also establish the initial risk that the decision-maker is accepting and calling on the CRM process to manage.

As discussed in Section 3.3.1, performance commitments are produced by the deliberators as a result of establishing risk tolerances on the performance measures. This facilitates deliberation of alternatives in terms of point value estimates of performance that reflect the deliberators' risk attitudes. The decision-maker may choose to keep the risk tolerances and performance commitments established by the deliberators, or he/she may choose to modify them in accordance with his/her own risk tolerances. In situations where the decision-maker's risk tolerances differ significantly from those established by the deliberators, the decision-maker may ask for additional deliberation in light of the modified commitments. In turn, the deliberators may ask the risk analysts for a revised risk list that reflects the new situation.

3.3.2.9 Documenting the Decision Rationale

The final step in the RIDM process is for the decision-maker to document the rationale for the selected alternative in the RISR. In a NASA program/project context, the RISR is developed in accordance with the activity's risk management plan. Information on documenting the decision rationale can be found in Appendix E, Content Guide for the Risk-Informed Selection Report.

Planetary Science Mission Example:
Deliberate, Select an Alternative, and Document the Decision Rationale

Deliberation and selection of an alternative is done in light of the TBfD which, in addition to the results presented in the previous Planetary Science Mission Example boxes, also contains the imposed constraint risk matrix.

Planetary Science Mission Imposed Constraint Risk Matrix

Alternative	Imposed Constraint Risk				
	Time to Completion	Project Cost	Data Volume	Planetary Contamination	Total*
	Constraint (< 55 months)	Constraint (<$500M)	Constraint (> 6 months)	Constraint (< 0.1% prob.)	
1. Propulsive Insertion, Low-Fidelity Science Package	2.8%	22%	4.1%	1.1%	25%
2. Propulsive Insertion, High-Fidelity Science Package	2.4%	57%	6.4%	3.2%	62%
3. Aerocapture, Low-Fidelity Science Package	3.0%	9.7%	8.7%	5.5%	18%
4. Aerocapture, High-Fidelity Science Package	2.3%	47%	12%	12%	57%

*This is the probability of failing to meet one or more of the imposed constraints. Because the performance measures are correlated, the total probability is not necessarily the sum of the individual imposed constraint risk probabilities. For example, if *time to completion* is greater than 55 months, then *data volume* is zero.

The first objective of the deliberators is to see whether or not the set of alternatives can be pruned down to a smaller set of contending alternatives to present to the decision maker. The imposed constraint risk matrix shows that the risk of not meeting the $500M cost constraint is high for Alternatives 2 and 4 compared to the agreed-upon risk tolerance of 27%. Specifically, Alternatives 2 and 4 are infeasible given the combination of the cost constraint and the JCL policy, which specifies that the project be budgeted at the 70th percentile or greater. The 70th percentile cost estimates are $860M for Alternative 2 and $650M for Alternative 4. Thus, the deliberators prune these alternatives from contention.

Alternatives 1 and 3 are identified as the contending alternatives that are recommended to the decision maker for consideration.

Planetary Science Mission Example:
Deliberate, Select an Alternative, and Document the Decision Rationale (continued)

In choosing between Alternatives 1 and 3, the decision maker must weigh differing performance capabilities with respect to competing objectives. Alternative 1, which uses propulsive insertion, has the following pros and cons:

Alternative 1: Propulsive Insertion, Low-Fidelity Science Package

Pros:

- A relatively low risk of not meeting the 6 month *data volume* imposed constraint

- The ability to commit to a higher data volume, given the decision-maker's risk tolerance

- A low probability of planetary contamination that is well within the decision-maker's risk tolerance

Cons:

- Higher cost, due to the need for a medium size launch vehicle

Conversely, Alternative 3 has the following pros and cons:

Alternative 3: Aerocapture, Low-Fidelity Science Package

Pros:

- Use of aerocapture technology, which the decision-maker considers to be a technology that promises future returns in terms of reduced payload masses for missions that can exploit aerocapture and/or aerobraking opportunities

- Lower cost, due to the use of a small launch vehicle afforded by the lower payload mass

Cons:

- Higher risk of not meeting the 6 month *data volume* imposed constraint, due to the potential for aerocapture failure during insertion

- Lower data volume at the decision-maker's risk tolerance

- A higher probability of planetary contamination, though still within the decision-maker's risk tolerance

The decision maker sees the choice in terms of whether or not the Planetary Science Mission is appropriate to use as an aerocapture test bed. If Alternative 3 is chosen and the mission succeeds, then not only will it advance the technology of aerocapture, but it will save money for this and future missions. If it fails, then the only near-term opportunity to gather important data on Planet "X" will be lost. It is a difficult decision, particularly because Alternative 3's *data volume* imposed constraint is near the edge of the decision-maker's risk tolerance. The decision-maker confers with selected deliberators and stakeholders, chooses the alternative that he believes best balances the pros and cons of each contending alternative, and documents his/her decision rationale in the RISR.

4.   SETTING THE STAGE FOR CRM[10]

As discussed earlier and illustrated in Figure 2, risk management is the combination of RIDM and CRM. The purpose of this handbook has been to provide guidance on implementing the RIDM requirements of NPR 8000.4A. Having done that in the first three sections, it is now time to set the stage for implementing the CRM requirements of the NPR. CRM should pick up where RIDM left off, informed by all that has been produced by the execution of the RIDM process at program/project onset.

4.1   RIDM "Initializes" CRM

As stated in Section 2.2, RIDM "initializes" CRM. What does it mean for RIDM to initialize CRM? The answer is that the RIDM process provides the CRM process with:

- A risk analysis for the selected alternative and other alternatives that were considered during the RIDM deliberation

- An initial set of identified and analyzed risks in the form of an initial risk list and associated analytical information

Note that in addition to the above, Systems Engineering must provide CRM with a set of risk management objectives in the form of a schedule of acceptable risk levels.

In accordance with NPR 8000.4A, CRM will manage implementation risks while "focused on the baseline performance requirements emerging from the RIDM process." After being "initialized," the job of CRM will be to complete the RIDM risk analysis (as described in Section 4.2.1) and to "burn-down" the risk of not achieving the performance requirements, to acceptable levels by the appropriate milestones. The starting point for the CRM process, or the initial risk levels for each of the performance measures, will be determined from the combination of 1) the performance requirements established by Systems Engineering and 2) the pdfs for each performance measure that come from the RIDM risk analysis (after it is completed within CRM). *Note: While the decision-maker's risk tolerance level for each performance measure established the values of the performance commitments, the performance commitments do not necessarily constitute the starting values of risk for the CRM process.*

The initial risk levels for each performance measure establish initial *risk acceptance levels*[11] for the achievement of performance requirements, with the expectation of improvement at key program/project milestones, and the objective of meeting the requirements per an associated *verification standard*[12]. In other words, as the program/project evolves over time, mitigations are implemented, and as risk concerns are retired and the state of knowledge about the performance measures improves, uncertainty should decrease; with an attendant lowering of residual risk (see Figure 37). The decrease may not be linear, as new risks may emerge during the project requiring

---

[10] This section is intended only to provide high-level concepts; details will follow in the forthcoming Risk Management Handbook.

[11] Risk acceptance levels are levels of risk considered to be acceptable by a decision maker at a given point in time.

[12] Verification standards are the standards used to verify that the performance requirements have been met.

new mitigations to be instituted. Overall, however, the tendency should be toward lower risk as time progresses.

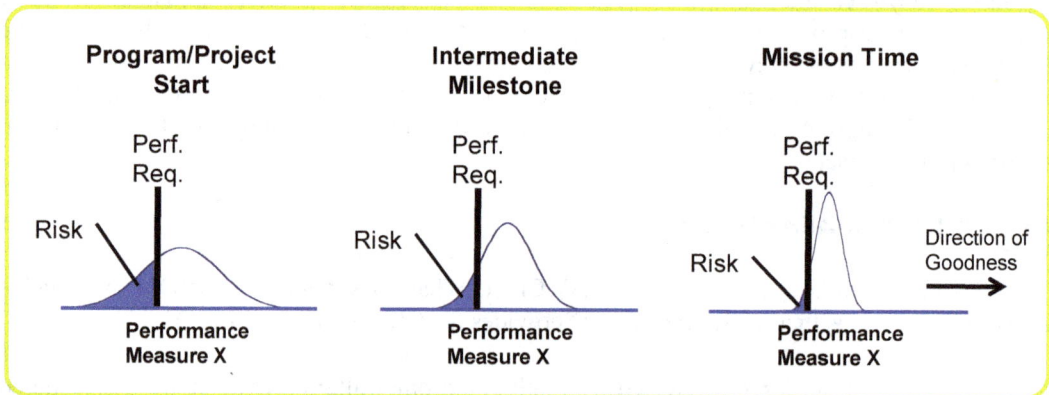

Figure 37. Decreasing Uncertainty and Risk over Time

4.2   Outputs of RIDM that Input to CRM

The following outputs of the RIDM process constitute inputs to the CRM process:

- A scenario-based risk analysis (Section 3.2) of the selected alternative and other alternatives developed down to the level of quantifiable causes for discriminator performance measures and to some lesser level of detail for the non-discriminator performance measures. The risk analysis will include:

  o   A list of the identified risk-driving uncertainties

  o   A working risk model(s) that include(s):

    ▪   Risk scenarios

    ▪   First-order mitigations

    ▪   Performance parameter distributions

    ▪   Performance measure distributions with imposed constraints, risk tolerance level, and performance commitments

- The Risk-Informed Selection Report, including the selected alternative and the initial performance commitment-based risk list (Figure 36) consisting of risks that impact discriminator, as well as non-discriminator performance measures, together with their associated risk statements and descriptive narratives.

4.2.1   Completing the RIDM Risk Analysis

The Risk Analysis developed as part of the RIDM process will have addressed all of the performance measures for the selected alternative, but only those performance measures considered discriminators among the alternatives will have been analyzed in detail. Because the initial risk list is based on the RIDM risk analysis, it is likely to contain only the major, top-level, initially evident risks and may therefore be incomplete, especially with respect to the non-discriminator performance measures.

As soon as feasible, the CRM process will need to complete the RIDM risk analysis for the non-discriminator performance measures and expand and update the initial risk list to include any new risks from the completed risk analysis. After the establishment of performance requirements by Systems Engineering, the risk analysis will have to be updated to reflect the performance requirements and resulting risks relative to them. Over the remaining life of the program/project, the risk list will also have to be updated with 1) the addition of new risks that become identifiable as the project progresses and 2) any changes that may be needed as risks evolve over time.

4.3   Output from Systems Engineering that Inputs to CRM

The output from Systems Engineering that constitutes input to the CRM process consists of the performance requirements for the selected alternative. The performance requirements developed by Systems Engineering are risk-informed, based on RIDM's performance measure pdfs, imposed constraints, risk tolerance levels and associated performance commitments, the risk analysis of the selected alternative, and any other information deemed pertinent by the Systems Engineering decision-maker.

Because the performance requirements are risk-informed but developed outside of the RIDM process, it is possible that their values may differ significantly from the performance commitments. In some cases the performance requirements may result in significantly higher risk levels compared to the performance commitments. *Note: In such cases it may be prudent for program/project management to negotiate new performance requirements with System Engineering.*

Since for the above reasons, the performance requirement values may differ from the corresponding performance commitment values of the selected alternative, the RIDM analysis process needs to check how any such difference translates into initial program risk acceptance levels. This is done by comparing the performance requirement values with the performance measure pdfs that were initially used to establish the performance commitments. The initial risk acceptance levels corresponding to the established performance requirements are transmitted to CRM, together with a schedule for their "burn-down."

4.4   Performing CRM

The purpose of CRM is to track whether the risk of not satisfying the performance requirements is being burned down in accordance with the progressively reduced limits on the risk, and to introduce mitigations when needed to satisfy the burn-down schedule. The burn-down schedule is to be referenced to and satisfied at certain program-selected milestones (e.g., SRR, SDR, PDR, CDR, etc.).

It is permissible in some cases for CRM to track performance margins as a surrogate for tracking risks. This is typically the case when the risk of not meeting a performance requirement is sufficiently low that it is not necessary to perform a detailed analysis of the uncertainties to verify that the burn-down profile is being satisfied.

4.5   The Continuing "Two-Way Street" Between RIDM and CRM

The flow of information between the RIDM and CRM processes begins in earnest, but should not end after alternative selection. Initially, RIDM (and Systems Engineering) will provide to CRM:

- Risks

- Risk Analysis

- Mitigations

- (From Systems Engineering) Initial performance requirement risk acceptance levels and "burn-down" schedule

Later, CRM may feedback to RIDM:

- Completed, updated risk analysis that includes new risks

- New risks for which mitigations are not available, as they arise

- New and revised risk mitigations, as they are developed

- Risks that demonstrate they cannot be controlled, as they arise

- New performance measures arising due to the addition of new performance requirements

- Verification that performance requirement risk burn-down objectives are being met at program-selected milestones

5.   REFERENCES

1.   NASA. *NPR 8000.4A, Agency Risk Management Procedural Requirements.* Washington, DC. 2008.

2.   NASA. *NASA/SP-2007-6105, NASA Systems Engineering Handbook.* Washington, DC. 2007.

3.   NASA. *NPR 7123.1, Systems Engineering Processes and Requirements.* Washington, DC. 2007.

4.   NASA. *NPR 7120.5D, NASA Space Flight Program and Project Management Processes and Requirements.* Washington, DC. 2007.

5.   NASA. *NPR 7120.7, NASA Information Technology and Institutional Infrastructure Program and Project Management Requirements*, Washington, DC. 2008.

6.   NASA. *NPR 7120.8, NASA Research and Technology Program and Project Management Requirements.* Washington, DC. 2008.

7.   NASA. *NPD 1000.5, Policy for NASA Acquisition.* Washington, DC. 2009.

8.   NASA. *NPD 7120.4C, Program/Project Management. Washington*, DC. 1999.

9.   NASA. *NPD 8700.1, NASA Policy for Safety and Mission Success.* Washington, DC. 2008.

10. Dezfuli, H., Stamatelatos, M., Maggio, G., and Everett, C., "Risk-informed Decision Making in the Context of NASA Risk Management," PSAM 10, Seattle, WA, 2010.

11. Hammond, J., Keeney, R., and Raiffa, H. "The Hidden Traps in Decision Making." *Harvard Business Review*, September – October 1998.

12. Clemen, R., *Making Hard Decisions.* Pacific Grove, CA. Duxbury Press, 1996.

13. Keeney, R., and Raiffa, H., *Decisions with Multiple Objectives: Preferences and Value Tradeoffs.* Cambridge, UK: Cambridge University Press, 1993.

14. Hammond, J., Keeney, R., and Raiffa, H., "Even Swaps: A Rational Method for Making Trade-offs." *Harvard Business Review*. March – April 1998.

15. U.S. Forest Service, Pacific Southwest Research Station. *Comparative Risk Assessment Framework and Tools (CRAFT)*, Version 1.0, 2005. (http://www.fs.fed.us/psw/topics/fire_science/craft/craft/index.htm)

16.   NASA Aerospace Safety Advisory Panel. "Aerospace Safety Advisory Panel Annual Report for 2009," Washington, D.C., 2010.

17.   NASA. "Constellation Program Implementation of Human-Rating Requirements," Tracking Number 2009-01-02a, Office of the Administrator, Washington, D.C., 2010.

18.   Keeney, R., *Value-Focused Thinking: A Path to Creative Decisionmaking*, Harvard University Press, 1992.

19.   Keeney, R., and McDaniels, T., "A Framework to Guide Thinking and Analysis Regarding Climate Change Policies," *Risk Analysis* 21, No. 6, pp. 989-1000, Society for Risk Analysis, 2001.

20.   NASA. "Exploration Systems Architecture Study -- Final Report," NASA-TM-2005-214062, Washington, DC. 2005.

21.   Maggio, G., Torres, A., Keisner, A., and Bowman, T., "Streamlined Process for Assessment of Conceptual Exploration Architectures for Informed Design (SPACE-AID)," Orlando, FL. 2005.

22.   NASA. *NPR 8715.3C, NASA General Safety Program Requirements.* Washington, DC. 2008.

23.   NASA. *NASA Cost Estimating Handbook.* Washington, DC. 2008.

24.   NASA. *Probabilistic Risk Assessment Procedures Guide for NASA Managers and Practitioners*, Version 1.1. Washington, DC. 2002.

25.   Morgan, M., and Henrion, M., *Uncertainty: A Guide to Dealing with Uncertainty in Quantitative Risk and Policy Analysis*, Cambridge Press, 1990.

26.   Apostolakis, G., "The Distinction between Aleatory and Epistemic Uncertainties is Important: An Example from the Inclusion of Aging Effects into Probabilistic Safety Assessment," Washington, DC. 1999.

27.   NASA. *NASA/SP-2009-569, Bayesian Inference for NASA Probabilistic Risk and Reliability Analysis.* Washington, DC. 2009.

28.   Daneshkhah, A., "Uncertainty in Probabilistic Risk Assessment: A Review," University of Sheffield Working Paper, August, 2004.

29.   Mosleh, A., Siu, N., Smidts, C., and Lui, C., "Model Uncertainty: Its Characterization and Quantification," University of Maryland, 1993.

30.   Attoh-Okine, N., and Ayyub, B., "Applied Research in Uncertainty Modeling and Analysis," International Series in Intelligent Technologies, Springer, 2005.

31. NASA. *NASA-STD-7009, Standard for Models and Simulations*. Washington, DC. 2008.

32. Groen, F., and Vesely, B., "Treatment of Uncertainties in the Comparison of Design Option Safety Attributes," PSAM 10, Seattle, WA, 2010.

33. NRC. *Understanding Risk – Informing Decisions in a Democratic Society.* The National Academies Press, 1996.

34. SEI. *Continuous Risk Management Guidebook.* 1996.

35. NASA. *NPD 1000.0A, Governance and Strategic Management Handbook.* Washington, DC. 2008.

36. Stromgren, C., Cates, G., and Cirillo, W., "Launch Order, Launch Separation, and Loiter in the Constellation 1½-Launch Solution," SAIC/NASA LRC, 2009.

37. Bearden, D., Hart, M., Bitten, R., et al, "Hubble Space Telescope (HST) Servicing Analyses of Alternatives (AoA) Final Report," The Aerospace Corporation, 2004.

APPENDIX A: NPR 8000.4A REQUIREMENTS CROSS-REFERENCE

Table A-1 provides a cross reference between RIDM-related requirements and good practices in NPR 8000.4A, and the RIDM Handbook sections that address them.

Table A-1. Cross-Reference Between RIDM-Related NPR 8000.4A "Good Practices" and Requirements, and RIDM Handbook Guidance[13]

NPR 8000.4A "Good Practices" and Requirements	
"Good Practices" relating to RIDM	**RIDM Handbook Guidance**
1.2.2.a. RIDM within each organizational unit involves:	
1.2.2.a. (1) Identification of decision alternatives, recognizing opportunities where they arise, and considering a sufficient number and diversity of performance measures to constitute a comprehensive set for decision-making purposes.	Section 3.1, Part 1 – Identification of Alternatives
1.2.2.a. (2) Risk analysis of decision alternatives to support ranking.	Section 3.2, Part 2 – Risk Analysis of Alternatives
1.2.2.a. (3) Selection of a decision alternative informed by (not solely based on) risk analysis results.	Section 3.3, Part 3 – Risk-Informed Alternative Selection
1.2.2.c. As part of a risk-informed process, the complete set of performance measure values (and corresponding assessed risks) is used, along with other considerations, within a *deliberative* process to improve the basis for decision making.	a) <u>Performance measure values</u>: Section 3.1.1.2, Derive Performance Measures b) <u>Deliberative process</u>: Section 3.3.2, Step 6 – Deliberate, Select an Alternative, and Document the Decision Rationale
1.2.2.d. Once a decision alternative has been selected for implementation, the performance measure values that informed its selection define the baseline performance requirements for CRM.	a) Section 2, RIDM Process Interfaces b) Section 4, Setting the Stage for CRM
1.2.2.e. ...for some purposes, decision making needs to be supported by quantification of the "aggregate risk" associated with a given performance measure; i.e., aggregation of all contributions to the risk associated with that performance measure.	Section 3.2.2, Step 4 – Conduct the Risk Analysis and Document the Results
1.2.2.e. ...the feasibility of quantifying aggregate risk is determined for each performance measure and then documented in the Risk Management Plan for each organizational unit.	Section 3.2.1, Step 3 – Set the Framework and Choose the Analysis Methodologies
1.2.4.a. RIDM and CRM [are] complementary processes that operate within every organizational unit.	Section 2.2, Coordination of RIDM and CRM
1.2.4.a. Each unit applies the RIDM process to decide how to fulfill its performance requirements and applies the CRM process to manage risks associated with implementation.	Section 2.2, Coordination of RIDM and CRM

---

[13] The "Good Practices" listed in Table A-1 are found in the indicated paragraphs of NPR 8000.4A. While they are not actual requirements, mandatory or advisory, they are the preferred ways to implement applicable parts of the NASA risk management process described in NPR 8000.4A.

NPR 8000.4A "Good Practices" and Requirements	
"Good Practices" relating to RIDM	**RIDM Handbook Guidance**
1.2.4.e. A basis for the evaluation of the performance measures and their associated risks should be agreed upon and documented in advance (or indicated by reference) in the Risk Management Plan.	a) Section 3.1.1.2, Derive Performance Measures b) Section 3.2.1, Step 3 – Set the Framework and Choose the Analysis Methodologies
1.2.4.f. It is the responsibility of the organizational unit at the higher level to assure that the performance requirements assigned to the organizational unit at the lower level reflect appropriate tradeoffs between/among competing objectives and risks.	a) Section 2.1, Negotiating Objectives across Organizational Unit Boundaries b) Section 3.3.2, Step – 6, Deliberate, Select an Alternative, and Document the Decision Rationale
1.2.4.f. The performance requirements can be changed, if necessary, but redefining and rebaselining them need to be negotiated with higher levels, documented, and subject to configuration control.	Section 2.2, Coordination of RIDM and CRM
1.2.4.g. Both CRM and RIDM are applied within a graded approach. The resources and depth of analysis need to be commensurate with the stakes and the complexity of the decision situations being addressed.	a) Section 3.2.1, Step 3 – Set the Framework and Choose the Analysis Methodologies b) Section 3.2.2, Step 4 – Conduct the Risk Analysis and Document the Results
3.3.2.2.a. Note: *The requirement to consider uncertainty is to be implemented in a graded fashion. If uncertainty can be shown to be small based on a simplified (e.g., bounding) analysis, and point estimates of performance measures clearly imply a decision that new information would not change, then detailed uncertainty analysis is unnecessary. Otherwise, some uncertainty analysis is needed to determine whether the expected benefit of the decision is affected significantly by uncertainty. In some cases, it may be beneficial to obtain new evidence to reduce uncertainty, depending on the stakes associated with the decision, the resources needed to reduce uncertainty, and programmatic constraints on uncertainty reduction activities (such as schedule constraints).*	Section 3.2.2, Step 4 – Conduct the Risk Analysis and Document the Results

NPR 8000.4A "Good Practices" and Requirements	
Requirements relating to RIDM	**RIDM Handbook Guidance**
3.1.1. The manager of each organizational unit shall:	
3.1.1.a. Ensure that the RIDM and CRM processes are implemented within the unit.	Section 2, RIDM Process Interfaces
3.1.1.d. Ensure that key decisions of the organizational unit are risk-informed. *Note Examples of key decisions include Architecture and design decisions, make-buy decisions, source selection in major procurements, budget reallocation (allocation of reserves).*	Section 1.4, When is RIDM Invoked?
3.1.1.e. Ensure that risks are identified and analyzed in relation to the performance requirements for each acquisition of the organizational unit and risk analysis results are used to inform the source selection.	a) Section 1.4, When is RIDM Invoked? b) Section 3.2.2, Step 4 – Conduct the Risk Analysis and Document the Results
3.1.1.i. Ensure that dissenting opinions arising during risk management decision making are handled through the dissenting opinion process as defined in NPR 7120.5D.	Section 3.2.2, Step 4 – Conduct the Risk Analysis and Document the Results
3.1.2 The risk manager of each organizational unit shall:	
3.1.2.a. Facilitate the implementation of RIDM and CRM.	RIDM Handbook
3.1.2.c.(2) Ensure the development of a Risk Management Plan that explicitly addresses safety, technical, cost, and schedule risks.	Section 3.1.1.2, Derive Performance Measures
3.1.2.c.(3) Ensure the development of a Risk Management Plan that delineates the organizational unit's approach for applying RIDM and CRM within a graded approach. *Note A "graded approach" applies risk management processes at a level of detail and rigor that adds value without unnecessary expenditure of unit resources.*	Section 3.2.1, Step 3 – Set the Framework and Choose the Analysis Methodologies
3.1.2.c.(4) Ensure the development of a Risk Management Plan that for each performance requirement, documents, or indicates by reference, whether its associated risks (including the aggregate risk) are to be assessed quantitatively or qualitatively and provides a rationale for cases where it is only feasible to assess the risk qualitatively.	a) Section 3.2.1, Step 3 – Set the Framework and Choose the Analysis Methodologies b) Section 3.2.2, Step 4 – Conduct the Risk Analysis and Document the Results
3.1.2.c.(6) Ensure the development of a Risk Management Plan that identifies stakeholders, such as Risk Review Boards, to participate in deliberations regarding the disposition of risks.	Section 3.3.1, Step 6 – Deliberate, Select an Alternative, and Document the Decision Rationale
3.1.2.c.(9) Ensure the development of a Risk Management Plan that delineates the processes for coordination of risk management activities and sharing of risk information with other affected organizational units.	Section 2, RIDM Process Interfaces

NPR 8000.4A "Good Practices" and Requirements	
Requirements relating to RIDM	**RIDM Handbook Guidance**
3.2 The manager of each organizational unit shall:	
3.2.a. The manager of each organizational unit shall ensure that performance measures defined for the organizational unit are used for risk analysis of decision alternatives to assist in RIDM.	a) Section 3.1.1.2, Derive Performance Measures b) Section 3.2.1, Step 3 – Set the Framework and Choose the Analysis Methodologies
3.2.b. The manager of each organizational unit shall ensure that the bases for performance requirement baselines (or rebaselines) are captured.	a) Section 3.3.1, Step 5 – Develop Risk-Normalized Performance Commitments b) Section 3.3.2, Step 6 – Deliberate, Select an Alternative, and Document the Decision Rationale
3.2.c. The manager of each organizational unit shall negotiate institutional support performance requirements with Center support units when required to meet program/project requirements.	Section 2.1, Negotiating Objectives across Organizational Unit Boundaries
3.2.d. The manager of each organizational unit shall ensure that performance measures defined for the organizational unit are used to scope the unit's CRM process (Requirement).	a) Section 2.2, Coordination of RIDM and CRM b) Section 4, Setting the Stage for CRM
3.3.2.1.b. The risk manager shall ensure that risk analyses performed to support RIDM are used as input to the "Identify" activity of CRM.	a) Section 2.2, Coordination of RIDM and CRM b) Section 4, Setting the Stage for CRM
3.3.2.2.a. The risk manager shall determine the protocols for estimation of the likelihood and magnitude of the consequence components of risks, including the timeframe, uncertainty characterization, and quantification when appropriate, and document these protocols in the Risk Management Plan.	a) Section 3.2.1, Step 3 – Set the Framework and Choose the Analysis Methodologies b) Section 3.2.2, Step 4 – Conduct the Risk Analysis and Document the Results
3.3.2.2.c. Wherever determined to be feasible (as documented in the Risk Management Plan), the risk manager shall ensure the characterization of aggregate risk through analysis (including uncertainty evaluation), as an input to the decision-making process.	a) Section 3.2.1, Step 3 – Set the Framework and Choose the Analysis Methodologies b) Section 3.2.2, Step 4 – Conduct the Risk Analysis and Document the Results

APPENDIX B: ACRONYMS AND ABBREVIATIONS

AHP	Analytic Hierarchy Process
AoA	Analysis of Alternatives
ASAP	Aerospace Safety Advisory Panel
CAS	Credibility Assessment Scale
CCDF	Complimentary Cumulative Distribution Function
CD	Center Director
CDR	Critical Design Review
CEV	Crew Exploration Vehicle
CFD	Computational Fluid Dynamics
Ci	Curie
CLV	Crew Launch Vehicle
ConOps	Concept of Operations
COS	Cosmic Origins Spectrograph
CRAFT	Comparative Risk Assessment Framework and Tools
CRM	Continuous Risk Management
DDT&E	Design, Development, Test & Evaluation
DM	De-orbit Module
DoD	Department of Defense
DRM	Design Reference Mission
EDS	Earth Departure Stage
EELV	Evolved Expendable Launch Vehicle
EOL	End of Life
EOM	End of Mission
ESAS	Exploration Systems Architecture Study
ETO	Earth-to-Orbit
FGS	Fine Guidance Sensor
FOM	Figure of Merit
G&A	General and Administrative
HQ	Headquarters
HST	Hubble Space Telescope
ISS	International Space Station
JCL	Joint Confidence Level
KSC	Kennedy Space Center
LCF	Latent Cancer Fatality

LEO	Low Earth Orbit
LLO	Low Lunar Orbit
LOC	Loss of Crew
LOI	Lunar Orbit Insertion
LOM	Loss of Mission
LSAM	Lunar Surface Access Module
M&S	Modeling & Simulation
MAUT	Multi-Attribute Utility Theory
MOE	Measure of Effectiveness
MSFC	Marshall Space Flight Center
MSO	Mission Support Office
NASA	National Aeronautics and Space Administration
NPD	NASA Policy Directive
NPR	NASA Procedural Requirements
NRC	Nuclear Regulatory Commission
NRC	National Research Council
ODC	Other Direct Cost
PC	Performance Commitment
pdf	Probability Density Function
PDR	Preliminary Design Review
PM	Performance Measure
pmf	Probability Mass Function
PnSL	Probability of No Second Launch
PR	Performance Requirement
PRA	Probabilistic Risk Assessment
RIDM	Risk-Informed Decision Making
RISR	Risk-Informed Selection Report
RM	Risk Management
RMP	Risk Management Plan
ROM	Rough Order-of-Magnitude
RRW	Risk Reduction Worth
RSRB	Redesigned Solid Rocket Booster
RTG	Radioisotope Thermoelectric Generator
S&MA	Safety & Mission Assurance
SDM	Service and De-orbit Module
SDR	System Design Review
SEI	Software Engineering Institute at Carnegie Mellon University
SM	Service Module
SME	Subject Matter Expert
SMn	Servicing Mission n

SRR	System Requirements Review
SSME	Space Shuttle Main Engine
STS	Space Transportation System
TBfD	Technical Basis for Deliberation
TLI	Trans-Lunar Injection
TRL	Technology Readiness Level
WBS	Work Breakdown Structure
WFC3	Wide Field Camera 3

APPENDIX C: DEFINITIONS

Aleatory: Pertaining to stochastic (non-deterministic) events, the outcome of which is described by a pdf. From the Latin alea (game of chance, die). [Adapted from [27]]

Consequence: The possible negative outcomes of the current conditions that are creating uncertainty. [Adapted from [34]]

Continuous Risk Management (CRM): A specific process for the management of risks associated with implementation of designs, plans, and processes. The CRM functions of identify, analyze, plan, track, control, and communicate and document provide a disciplined environment for continuously assessing what could go wrong, determining which issues are important to deal with, and implementing strategies for dealing with them. [Adapted from [34]]

Deliberation: Any process for communication and for raising and collectively considering issues. In deliberation, people discuss, ponder, exchange observations and views, reflect upon information and judgments concerning matters of mutual interest, and attempt to persuade each other. Deliberations about risk often include discussions of the role, subjects, methods, and results of analysis. [Excerpted from [33]]

Dominated Alternative: An alternative that is inferior to some other alternative with respect to every performance measure.

Epistemic: Pertaining to the degree of knowledge. From the Greek episteme (knowledge). [Adapted from [27]]

Imposed Constraint: A limit on the allowable values of the performance measure with which it is associated. Imposed constraints reflect performance requirements that are negotiated between NASA organizational units and which define the task to be performed.

Likelihood: Probability of occurrence.

Objective: A specific thing that you want to achieve. [12]

Performance Commitment: A level of performance that a decision alternative is intended to achieve at a given level of risk. Performance commitments are established on the performance measures of each alternative as a means of comparing performance across alternatives that is consistent with the alternative-independent risk tolerance of the decision-maker.

Performance Measure: A metric used to measure the extent to which a system, process, or activity fulfills its associated performance objective. [Adapted from [1]]

Performance Objective: An objective whose fulfillment is directly quantified by an associated performance measure. In the RIDM process, performance objectives are derived via an objectives hierarchy, and represent the objectives at the levels of the hierarchy.

Performance Parameter: Any value needed to execute the models that quantify the performance measures. Unlike performance measures, which are the same for all alternatives, performance parameters typically vary among alternatives, i.e., a performance parameter that is defined for one alternative might not apply to another alternative.

Performance Requirement: The value of a performance measure to be achieved by an organizational unit's work that has been agreed-upon to satisfy the needs of the next higher organizational level. [1]

Quantifiable: An objective is quantifiable if the degree to which it is satisfied can be represented numerically. Quantification may be the result of direct measurement; it may be the product of standardized analysis; or it may be assigned subjectively.

Risk: In the context of RIDM, risk is the potential for shortfalls, which may be realized in the future, with respect to achieving explicitly-stated performance commitments. The performance shortfalls may be related to institutional support for mission execution, or related to any one or more of the following mission execution domains: safety, technical, cost, schedule.

Risk is operationally defined as a set of triplets:
    a. The scenario(s) leading to degraded performance in one or more performance measures,
    b. The likelihood(s) of those scenarios,
    c. The consequence(s), impact, or severity of the impact on performance that would result if those scenarios were to occur.

Uncertainties are included in the evaluation of likelihoods and consequences. [Adapted from [1]]

Risk Analysis: For the purpose of this Handbook, risk analysis is defined as the probabilistic assessment of performance such that the probability of not meeting a particular performance commitment can be quantified.

Risk Averse: The risk attitude of preferring a definite outcome to an uncertain one having the same expected value.

Risk-Informed Decision Making: A risk-informed decision-making process uses a diverse set of performance measures (some of which are model-based risk metrics) along with other considerations within a deliberative process to inform decision making.

> *Note: A decision-making process relying primarily on a narrow set of model-based risk metrics would be considered "risk-based."* [1]

Risk Management: Risk management includes RIDM and CRM in an integrated framework. This is done in order to foster proactive risk management, to better inform decision making through better use of risk information, and then to more effectively manage implementation risks by focusing the CRM process on the baseline performance requirements emerging from the RIDM process. [1]

Risk Seeking: The risk attitude of preferring an uncertain outcome to a certain one having the same expected value.

Robust: A robust decision is one that is based on sufficient technical evidence and characterization of uncertainties to determine that the selected alternative best reflects decision-maker preferences and values given the state of knowledge at the time of the decision, and is considered insensitive to credible modeling perturbations and realistically foreseeable new information.

Scenario: A sequence of credible events that specifies the evolution of a system or process from a given state to a future state. In the context of risk management, scenarios are used to identify the ways in which a system or process in its current state can evolve to an undesirable state.

Sensitivity Study: The study of how the variation in the output of a model can be apportioned to different sources of variation in the model input and parameters. [31]

Stakeholder: A stakeholder is an individual or organization that is materially affected by the outcome of a decision or deliverable but is outside the organization doing the work or making the decision. [35]

Performance Parameter: Collectively, performance parameters specify how an alternative is going to accomplish its performance objectives and comply with imposed constraints. Unlike performance objectives, which are the same for all alternatives, performance parameters are unique to each alternative.

Uncertainty: An imperfect state of knowledge or a physical variability resulting from a variety of factors including, but not limited to, lack of knowledge, applicability of information, physical variation, randomness or stochastic behavior, indeterminacy, judgment, and approximation. [1]

APPENDIX D: CONTENT GUIDE FOR THE TECHNICAL BASIS FOR DELIBERATION

Technical Basis for Deliberation Content

The Technical Basis for Deliberation (TBfD) document is the foundation document for the risk-informing activities conducted during Part 1 and Part 2 of the RIDM Process. The TBfD conveys information on the performance measures and associated imposed constraints for the analyzed decision alternatives.

Because the TBfD provides the specific risk information to understand the uncertainty associated with each alternative, this document serves as the technical basis for risk-informed selection of alternatives within the program or project. The risk analysis team, working under the overall program/project guidance, develops TBfD documentation and updates the information provided as necessary based upon questions and/or concerns of stakeholders during deliberation. The risk analysis team works with the deliberators and decision-maker to support deliberation and alternative selection.

The TBfD includes the following general sections:

- Technical Summary: This section describes the problem to be solved by this effort and each of the general contexts of each of the alternatives.

- Top-level Requirements and Expectations: This section contains the top-level requirements and expectations identified in Step 1 of the RIDM process. In cases involving diverse stakeholders, a cross reference between expectations and stakeholder may be presented.

- Derivation of Performance Measures: This section shows the derivation of performance measures for the decision conducted in Step 1 of the RIDM process. Typical products are the objectives hierarchy and a table mapping the performance objectives to the performance measures. When proxy performance measures are used, their definitions are provided along with the rationale for their appropriateness. When constructed scales are used, the scales are presented.

- Decision Alternatives: This section shows the compilation of feasible decision alternatives conducted in Step 2 of the RIDM process. Typical products are trade trees, including discussion of tree scope and rationales for the pruning of alternatives prior to risk analysis. Alternatives that are retained for risk analysis are described. This section also identifies any imposed constraints on the allowable performance measure values, and a map to the originating top-level requirements and/or expectations.

- Risk Analysis Framework and Methods: This section presents the overall risk analysis framework and methods that are set in Step 3 of the RIDM process. For each analyzed alternative, it shows how discipline-specific models are integrated into an analysis process that preserves correlations among performance parameters. Discipline-specific analysis models are identified and rationale for their selection is given. Performance parameters are identified for each alternative.

- Risk Analysis Results: This section presents the risk analysis results that are quantified in Step 4 of the RIDM process.

  o Scenario descriptions: For each alternative, the main scenarios identified by the risk analysis are presented.

  o Performance measure pdfs: For each alternative, the marginal performance measure pdfs are presented, along with a discussion of any significant correlation between pdfs.

  o Imposed constraint risk: For each alternative, the risk with respect to imposed constraints is presented, along with a discussion of the significant drivers contributing to that risk.

  o Supporting analyses: For each alternative, uncertainty analyses and sensitivity studies are summarized.

- Risk Analysis Credibility Assessment: This sections presents the credibility assessment performed in accordance with [31].

APPENDIX E: CONTENT GUIDE FOR THE RISK-INFORMED SELECTION REPORT

Risk-Informed Selection Report Content

The Risk-Informed Selection Report (RISR) documents the rationale for selection of the selected alternative and demonstrates that the selection is risk-informed. The decision-maker, working with the deliberators and risk analysis team, develops the RISR.

The RISR includes the following general sections:

- Executive Summary: This summary describes the problem to be solved by this effort and each of the general contexts of each of the alternatives. It identifies the organizations and individuals involved in the decision-making process and summarizes the process itself, including any intermediate downselects. It presents the selected alternative and summarizes the basis for its selection.

- Technical Basis for Deliberation: This section contains material from the TBfD (see Appendix D).

- Performance Commitments: This section presents the performance measure ordering and risk tolerances used to develop the performance commitments during Step 5 of the RIDM process, with accompanying rationale. It tabulates the resultant performance commitments for each alternative.

- Deliberation: This section documents the issues that were deliberated during Step 6 of the RIDM process.

  - Organization of the deliberations: The deliberation and decision-making structure is summarized, including any downselect decisions and proxy decision-makers.

  - Identification of the contending decision alternatives: The contending alternatives are identified, and rationales given for their downselection relative to the pruned alternatives. Dissenting opinions are also included.

  - Pros and cons of each contending alternative: For each contending alternative, its pros and cons are presented, along with relevant deliberation issues including dissenting opinions. This includes identifying violations of significant engineering standards, and the extent to which their intents are met by other means.

  - Deliberation summary material: Briefing material, etc., from the deliberators and/or risk analysts to the decision-maker (or decision-makers, in the case of multiple downselects) is presented.

- Alternative Selection: This section documents the selection of an alternative conducted in Step 6 of the RIDM process.

o   Selected alternative: The selected alternative is identified, along with a summary of the rationale for its selection.

o   Performance commitments: The finalized performance commitments for the selected alternative are presented, along with the final performance measure risk tolerances and performance measure ordering used to derive them.

o   Risk list: The RIDM risk list for the selected alternative is presented, indicating the risk-significant conditions extant at the time of the analysis, and the assessed impact on the ability to meet the performance commitments.

o   Decision robustness: An assessment of the robustness of the decision is presented.

APPENDIX F: SELECTED NASA EXAMPLES OF RIDM PROCESS ELEMENTS

NASA has a long history of incorporating risk considerations into its decision-making processes. As part of the development of this handbook, NASA OSMA reviewed a number of decision forums and analyses for insights into the needs that the handbook should address, and for examples of decision-making techniques that are illustrative of elements of the resultant RIDM process.

The following example process elements are intended as illustrations of the general intent of the RIDM process elements to which they correspond. They do not necessarily adhere in every detail to the guidance in this handbook. Nevertheless, they represent sound techniques that have risk-informed decision making at NASA.

F.1 Stakeholder Expectations

F.1.1 The Use of Design Reference Missions in ESAS [20]

A series of DRMs was established to facilitate the derivation of requirements and the allocation of functionality between the major architecture elements. Three of the DRMs were for missions to the International Space Station (ISS): transportation of crew to and from the ISS, transportation of pressurized cargo to and from the ISS, and transportation of unpressurized cargo to the ISS. Three of the DRMs were for lunar missions: transportation of crew and cargo to and from anywhere on the lunar surface in support of 7-day "sortie" missions, transportation of crew and cargo to and from an outpost at the lunar south pole, and one-way transportation of cargo to anywhere on the lunar surface. A DRM was also established for transporting crew and cargo to and from the surface of Mars for an 18-month stay. Figures F-1 and F-2 show two of the ESAS DRMs: one for an ISS mission and one for a lunar mission.

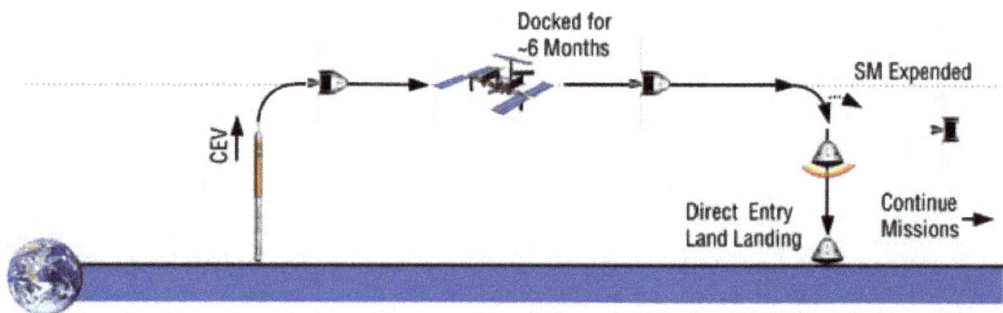

Figure F-1. Crew Transport to and from ISS DRM

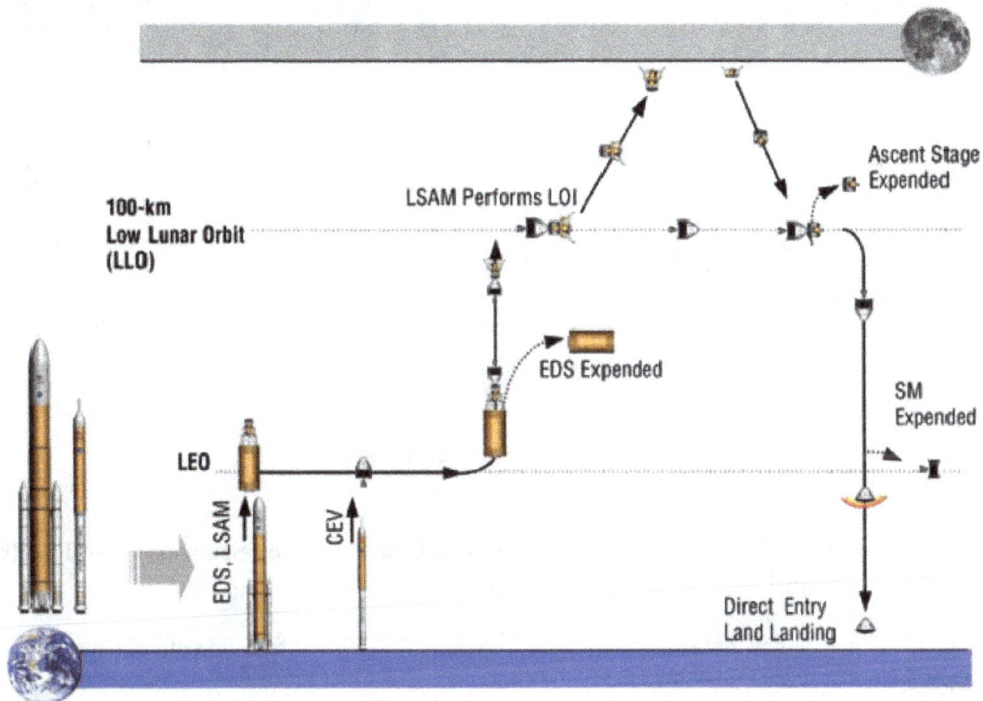

Figure F-2. Lunar Sortie Crew with Cargo DRM

F.2 Objectives Hierarchies and Performance Measures

F.2.1 The Use of Figures of Merit (FOMs) in ESAS [20]

The various trade studies conducted by the ESAS team used a common set of FOMs (a.k.a. performance measures) for evaluation.[14] Each option was quantitatively or qualitatively assessed against the FOMs shown in Figure F-3. FOMs were included in the areas of: safety and mission success, effectiveness and performance, extensibility and flexibility, programmatic risk, and affordability. FOMs were selected to be as mutually exclusive and measurable as possible.

---

[14] The inclusion of this example should not be taken as advocating any particular set of performance measures. In particular, the treatment of risk in terms of explicit risk FOMs is inconsistent with the RIDM process as discussed in Section 3.1.2.

Safety and Mission Success	Effectiveness/ Performance	Extensibility/ Flexibility	Programmatic Risk	Affordability
Probability of Loss of Crew (P(LOC))	Cargo Delivered to Lunar Surface	Lunar Mission Flexibility	Technology Development Risk	Technology Development Cost
Probability of Loss of Mission (P(LOM))	Cargo Returned from Lunar Surface	Mars Mission Extensibility	Cost Risk	Design, Development, Test, and Evaluation (DDT&E) Cost
	Surface Accessibility	Extensibility to Other Exploration Destinations	Schedule Risk	Facilities Cost
	Usable Surface Crew-Hours		Political Risk	Operations Cost
	System Availability	Commercial Extensibility		Cost of Failure
	System Operability	National Security Extensibility		

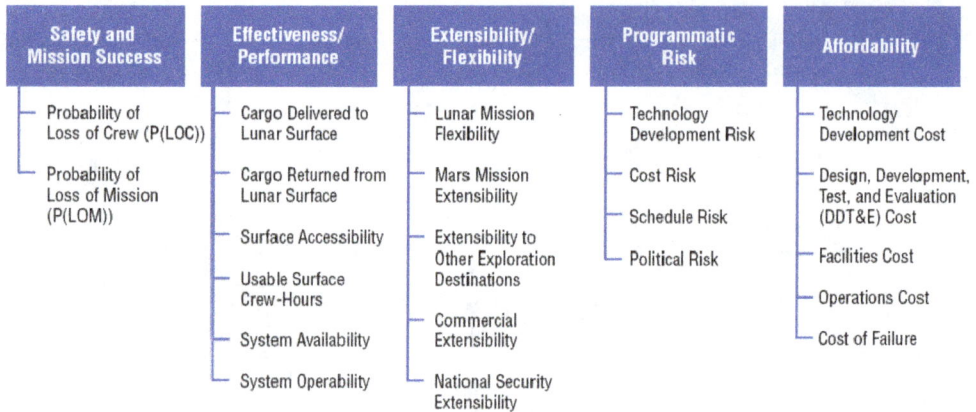

Figure F-3. ESAS FOMs

F.2.2   The Use of Figures of Merit in "Launch Order, Launch Separation, and Loiter in the Constellation 1½-Launch Solution" [36]

The goal of this launch order analysis was to evaluate the identified operational concepts and then produce a series of relevant FOMs for each one. The most basic metric that was considered was the probability that each concept would result in a failure to launch the second vehicle. The FOMs for the study had to cover a number of areas that were significant to decision-makers on selecting a concept. Table F-1 lists the FOMs that were considered in this study.

Table F-1. Launch Order Risk Analysis FOMs

FOMs
Probability of No Second Launch
Cost of Failure
Loss of Delivery Capability to the Surface
Additional Risk to the Crew
Additional Costs or Complexities

F.3 Compiling Alternatives

F.3.1 The Use of Trade Trees in ESAS [20]

Figure F-4 shows the broad trade space for Earth-to-orbit (ETO) transportation defined during ESAS. In order to arrive at a set of manageable trade options, external influences, as well as technical influences, were qualitatively considered in order to identify feasible decision alternatives.

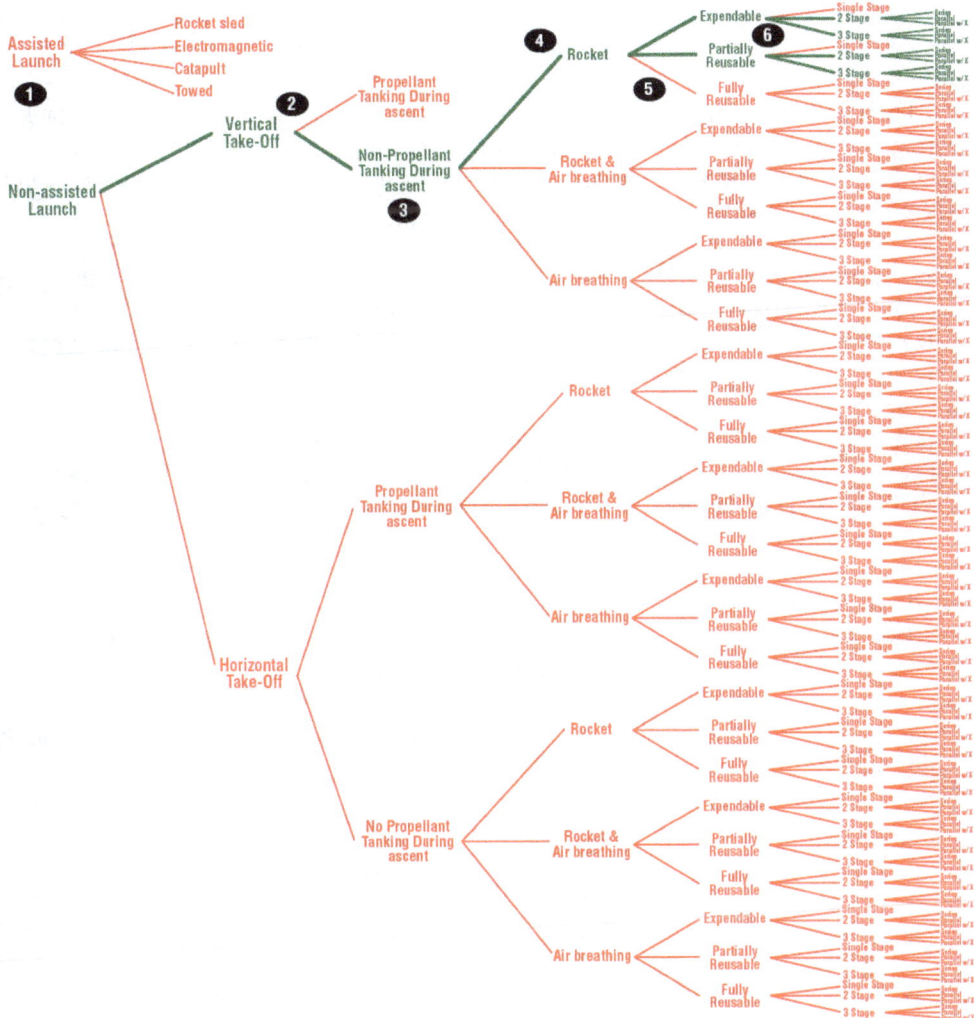

Figure F-4. Possible Range of ESAS Launch Trade Study

The decision points, illustrated numerically in Figure F-4, are described below, with the subsequent study decisions and supporting rationale.

- Non-assisted versus Assisted Takeoff: Assisted launch systems (e.g., rocket sled, electromagnetic sled, towed) on the scale necessary to meet the payload lift requirements are beyond the state-of-the-art for near-term application. Therefore, Non-assisted Takeoff was chosen.

- Vertical versus Horizontal Takeoff: Current horizontal takeoff vehicles and infrastructures are not capable of accommodating the gross takeoff weights of concepts needed to meet the payload lift requirements. Therefore, Vertical Takeoff was chosen.

- No Propellant Tanking versus Propellant Tanking During Ascent: Propellant tanking during vertical takeoff is precluded due to the short period of time spent in the atmosphere 1) to collect propellant or 2) to transfer propellant from another vehicle. Therefore, No Propellant Tanking was chosen.

- Rocket versus Air Breathing versus Rocket and Air Breathing: Air breathing and combined cycle (i.e., rocket and air breathing) propulsion systems are beyond the state-of-the-art for near-term application and likely cannot meet the lift requirements. Therefore, Rocket was chosen.

- Expendable versus Partially Reusable versus Fully Reusable: Fully reusable systems are not cost-effective for the low projected flight rates and large payloads. Near-term budget availability and the desire for a rapid development preclude fully reusable systems. Therefore, Expendable or Partially Reusable was chosen.

- Single-stage versus 2-Stage versus 3-Stage: Single-stage concepts on the scale necessary to meet the payload lift requirements are beyond the state-of-the-art for near-term application. Therefore, 2-Stage or 3-Stage was chosen.

- Clean-sheet versus Derivatives of Current Systems: Near-term budget availability and the desire for a rapid development preclude clean-sheet systems. Therefore, Derivatives of Current Systems was chosen.

F.3.2 The Use of a Trade Tree in "Launch Order, Launch Separation, and Loiter in the Constellation 1½-Launch Solution" [36]

The options that were evaluated in this decision analysis are depicted in the trade tree of Figure F-5. Two options were considered for launch order: launching Ares I first, followed by Ares V, identified as "I-V"; and launching Ares V first, followed by Ares I, identified as "V-I". In addition, two types of LEO loiter duration were considered. The first loiter option was to support only a single TLI window. The second loiter option was to support multiple TLI windows. Because of the limited loiter duration of the Orion crew module in LEO, the option to support multiple TLI windows is applicable only to a V-I launch order. Finally, options for the planned separation between the two launches of 90-minutes and 24-hours were evaluated. The ESAS baseline of a V-I launch order, a loiter duration that supports multiple TLI windows, and a launch separation of 24-hours is identified in Figure F-5.

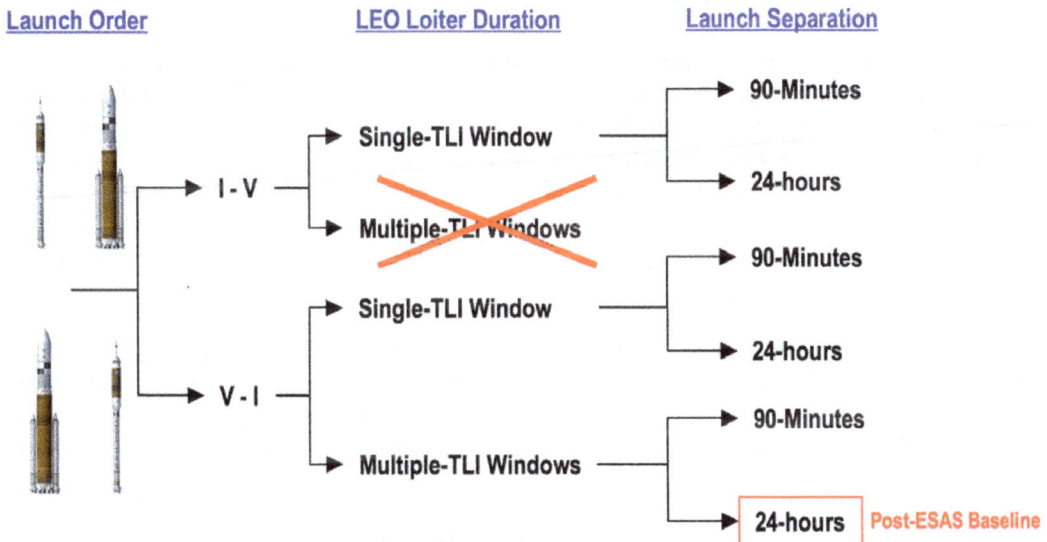

Figure F-5. Launch Order Analysis Trade Space

F.3.3 Hubble Space Telescope (HST) Servicing Analyses of Alternatives (AoA) [37]

Figure F-6 is the top-level robotic servicing decision tree that was used in the HST servicing analysis to scope out the space of mission concepts within which to compile specific alternatives. Due to a driving concern of an uncontrolled re-entry of HST, the expectation at NASA since program inception was that HST would be disposed of at end of life. Achieving NASA's casualty expectation standard of less than 1 in 10,000 requires some degree of active disposal of HST. Active disposal requires a minimum capability to rendezvous and dock with HST, and then either to boost the observatory to a disposal orbit or perform a controlled re-entry.

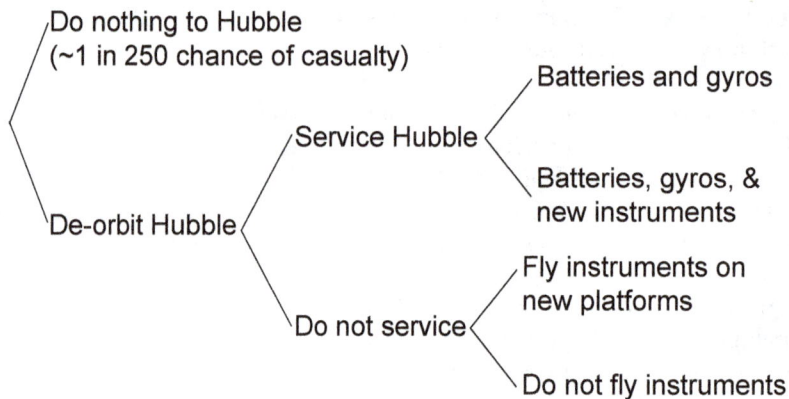

Figure F-6. Robotic Servicing Decision Tree

The alternative development began in brainstorming sessions in which the study team developed "clean-sheet" concepts, which encompassed doing nothing to HST, rehosting the SM4 instruments on new platforms, robotic servicing, and astronaut servicing. The brainstorming approach involved capturing the full set of ideas suggested by a large group of study team members during several alternatives development meetings. These ideas and concepts were then grouped into broad categories. The resulting database was augmented by internet and literature searches for related ideas, including the private sector responses on robotic servicing approaches and technologies.

Table F-2 provides the full set of alternatives, from which the final set was selected. A number of unorthodox options, including foreign participation in the development of a mission, delivery of ordnance and detonation just prior to the re-entry atmospheric interface, and forcing breakup at re-entry with a missile defense interceptor were dismissed as being too risky, as well as politically and/or technically infeasible.

Table F-2. Alternatives Brainstorming.

Classification	ID	Description
Minimum Action	1	Do nothing
Minimum Action	2	Extend life through ground-based operational workarounds
Minimum Action	3	Use HST attitude modulation to control entry point
Rehost Option	4	Fly COS, WFC3, and FGS on new platform
Rehost Option	5	Fly COS, WFC3, and FGS replacement instruments on new platform
Rehost Option	6	Fly HST current and/or SM4 instruments on platforms already in development
Rehost Option	7	Fly HST current and/or SM4 replacement instruments on platforms already in development
Rehost Option	8	Rebuild HST
Rehost Option	9	Replace full HST capability on a new platform
De-orbit	10	Dock before EOM, de-orbit immediately

De-orbit	11	Dock before EOM, continue mission and de-orbit after EOM
De-orbit	12	Dock after EOM, but before EOL and de-orbit immediately
De-orbit	13	Dock after EOL and de-orbit
Service & De-orbit	14	Service-only mission (separate de-orbit mission)
Service & De-orbit	15	Launch SDM, service, stay attached for de-orbit
Service & De-orbit	16	Launch SDM, service, detach and station-keep, then reattach for de-orbit
Service & De-orbit	17	Launch SM and DM together, service, remove SM, DM stays on
Robotic Option	18	De-orbit vs. graveyard orbit
Robotic Option	19	Propulsive vs. attitude modulation reentry
Robotic Option	20	External attachment vs. internal replacement vs. internal attachment
Robotic Option	21	Autonomous vs. telerobotic docking
Robotic Option	22	Reboost as part of all servicing missions
Robotic Option	23	Timing of docking and de-orbiting (before/after EOM/EOL)
Robotic Option	24	Scope of servicing: full vs. extended life only (batteries + gyros + instruments vs. batteries + gyros)
Unconventional	25	De-orbit via foreign partners or commercial firms
Unconventional	26	Service via foreign partners or commercial firms
Unconventional	27	CEV crewed servicing mission
Unconventional	28	Uncontrolled de-orbit + KKV forced breakup at atmospheric interface
Unconventional	29	Uncontrolled de-orbit + detonation forced breakup at atmospheric interface
Unconventional	30	Use tug to deliver HST to ISS orbit for service by STS crew
Unconventional	31	Propulsively transfer to ISS and retrieve via shuttle
Unconventional	32	Point existing Earth-viewing orbital instruments upwards
Unconventional	33	Safe haven near HST for Shuttle servicing
Astronaut Servicing	34	Do SM4 (on STS)

A natural grouping emerged from this set of alternatives, whereby several complementary alternatives could be condensed into a single concept. Brainstorming was next linked with a deductive method. Options were distilled into four general families: Rehost, Disposal, Service, and Safe Haven as shown in Figure F-7.

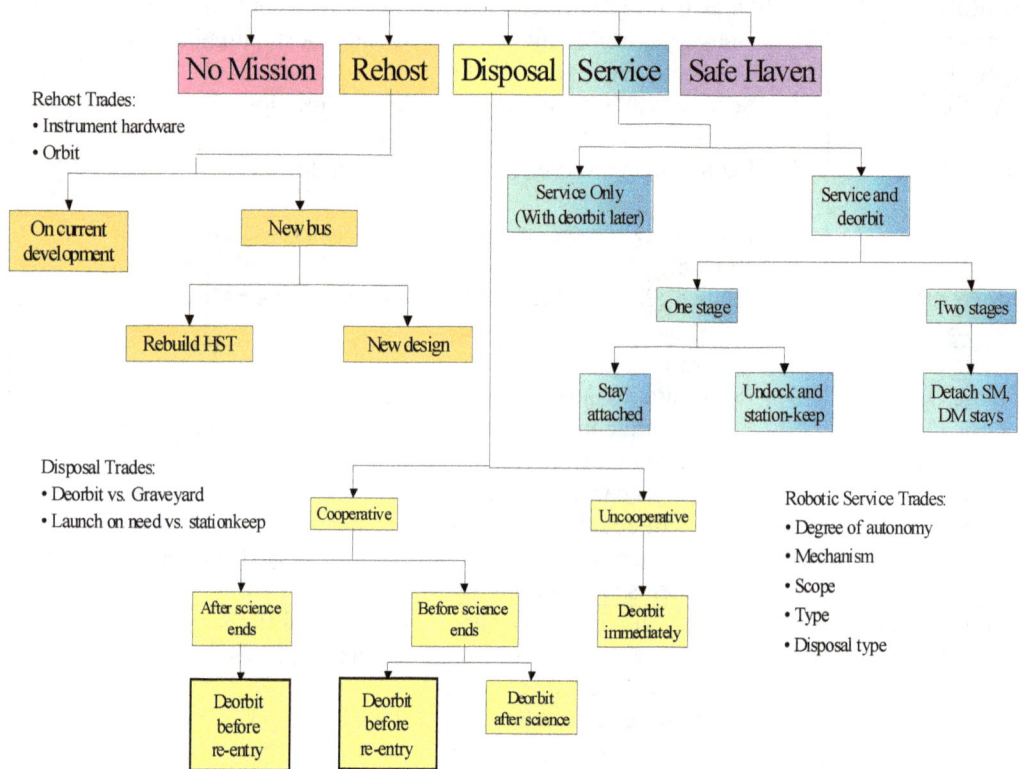

Figure F-7. Option Tree Analysis

The study team used a methodical approach to the process of selecting and constructing the final set of alternatives to ensure that the important elements in the robotic servicing trade space were included. The branches of the option tree were developed to a low enough level to cover various architectural and technology options at the conceptual level, such as the type of robotic mechanism and the amount of servicing performed.

F.4 Risk Analysis of Alternatives

F.4.1 Scenario-Based, Probabilistic Analysis of Performance in ESAS [20]

One of the issues analyzed in ESAS was that of monostability of the entry vehicle. In this context, a monostable entry vehicle will only aerodynamically trim in one attitude, such that the vehicle would always be properly oriented for entry (similar to Soyuz). Requiring an entry to be inherently monostable results in an outer mold line with weight and packaging issues. ESAS looked at how much benefit, from a crew safety risk (i.e., P(LOC)) standpoint, monostability provides, so that the costs can be traded within the system design. In addition, ESAS looked at additional entry vehicle systems that are required to realize the benefits of monostability and considered systems that could remove the need to be monostable.

The risk analysis consisted of two parts: a flight mechanics stability element and a P(LOC) assessment. The two pieces were combined to analyze the risk impact of entry vehicle stability. The risk assessment was performed using the simple event tree shown in Figure F-8, representing the pivotal events during the entry mission phase. Each pivotal event was assigned a success probability determined from historical reliability data. In addition, mitigations to key pivotal events were modeled using the results from the stability study, as were the success probabilities for ballistic entry. In the event tree, the "Perform Ballistic Entry" event mitigates the "Perform Entry" (attitude and control) event, while the "Land and Recover from Ballistic Entry" event replaces the "Land and Recover" event should a ballistic entry occur.

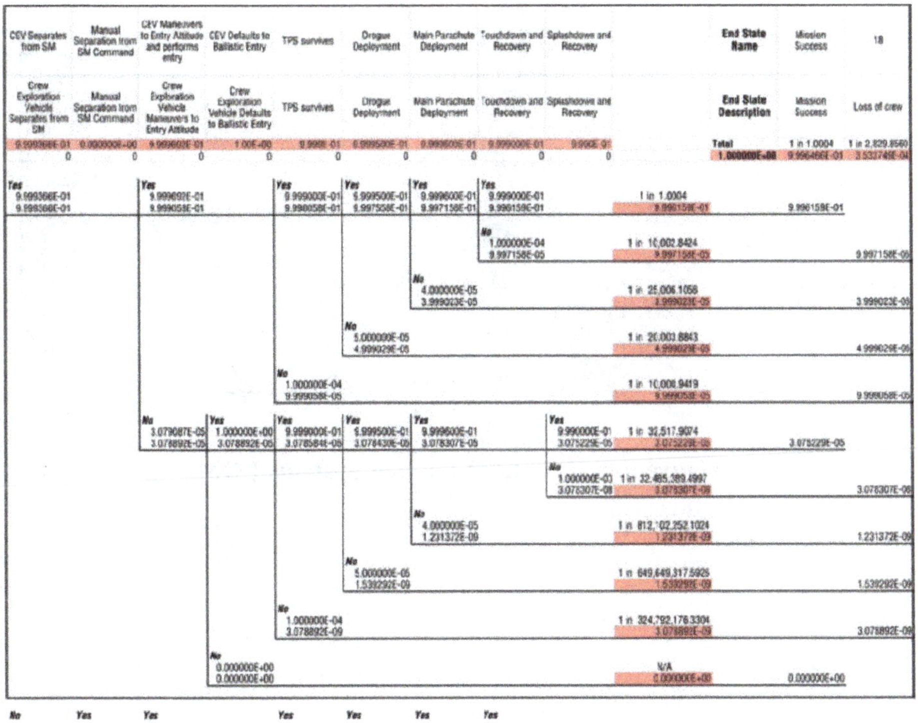

Figure F-8. ESAS Entry, Descent, and Landing Event Tree

F.4.2 Probabilistic Analysis of CLV Crew Safety Performance and Mission Success Performance in ESAS [20]

ESAS assessed more than 30 launch vehicle concepts to determine P(LOM) and P(LOC) estimates. Evaluations were based on preliminary vehicle descriptions that included propulsion elements and Shuttle-based launch vehicle subsystems. The P(LOM) and P(LOC) results for each of the CLV results are shown graphically in Figures F-9 and F-10, respectively. The results are expressed as probability distributions over the epistemic uncertainty modeled in the analysis, indicating the range of possible values for P(LOM) and P(LOC) given the state of knowledge at the time the analysis was done. Aleatory uncertainty has been accounted for in the analysis by expressing the results in terms of the probabilistic performance measures of P(LOM) and P(LOC). These measures represent the expected values for loss of mission and loss of crew, respectively.

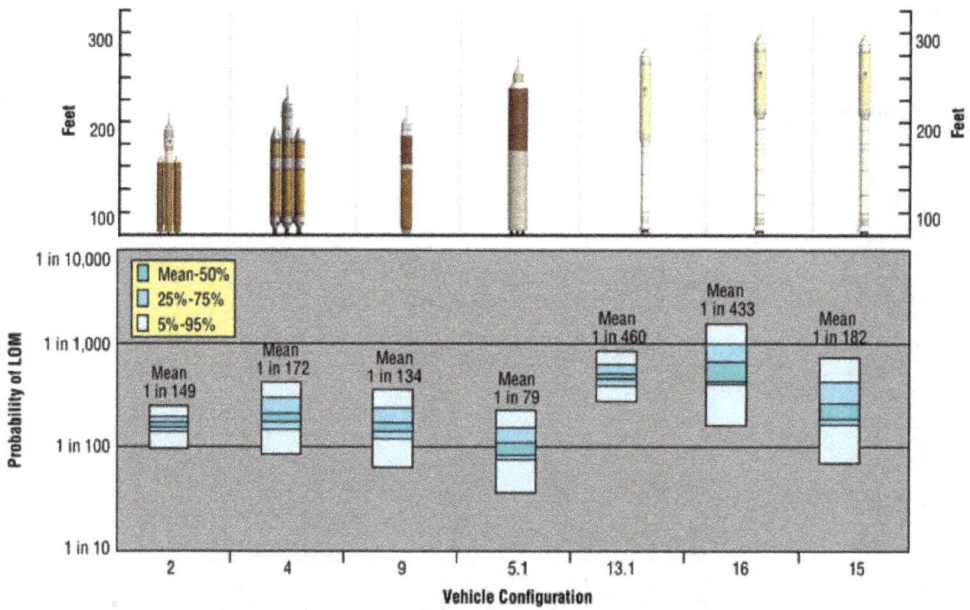

Figure F-9. CLV LEO Launch Systems LOM

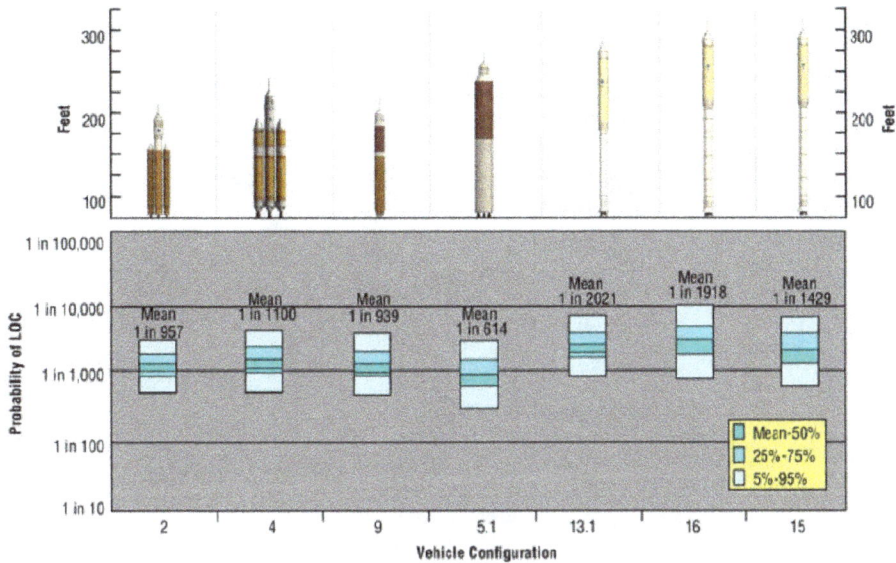

Figure F-10. CLV LEO Launch Systems LOC

F.4.3 Downselection in "Launch Order, Launch Separation, and Loiter in the Constellation 1½-Launch Solution" [36]

The launch order analysis down-selected to the preferred option through various down-select cycles that sequentially pruned options from the trade tree by focusing on various FOMs in each down-selection cycle until only one branch was left. Figure F-11 shows the overall trade tree and the down-selections made through 4 iterations considering various FOMs in pruning the tree, including a summary of each down-select rationale. The first down-select eliminating multiple TLI window support for a I-V launch order was based on the trade space constraint of not modifying the current Orion vehicle, which has a capacity to loiter in LEO for a maximum of four days, limited by consumables.

Launch Order **LEO Loiter Duration** **Launch Separation**

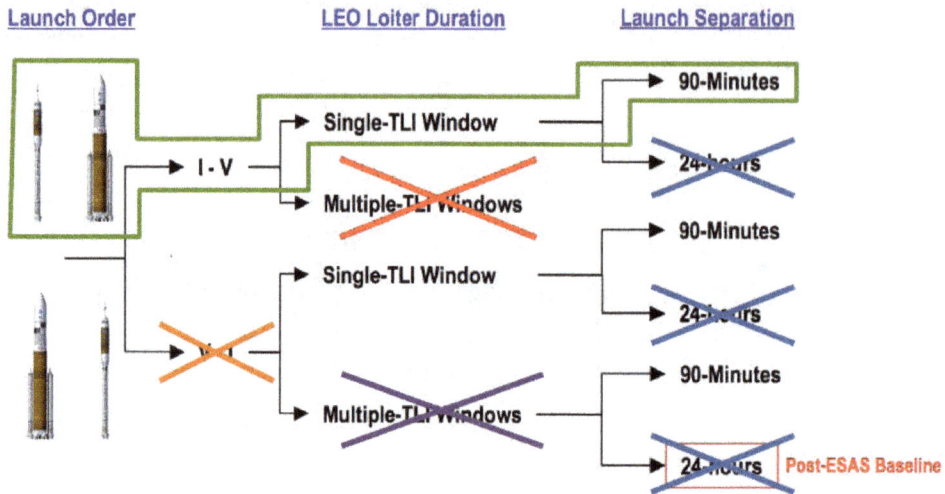

Single-TLI Window — 90-Minutes

I - V

Multiple-TLI Windows — 24-hours

Single-TLI Window — 90-Minutes

V - I — 24-hours

Multiple-TLI Windows — 90-Minutes

24-hours Post-ESAS Baseline

1. Option not viable since crew would be in LEO too long
2. Options eliminated due to much higher PnSL
3. Options eliminated due to the high DDT&E costs and increased development risk of baselining a higher performance Ares V (Single-TLI Window LEO Loiter set to 4 days at this point since anything less would sub-optimal from a replacement cost perspective)
4. Options eliminated due to higher expected replacement costs (based on Ares V being 2.88 times more expensive than Ares I)

Option selected (4 day LEO Loiter providing a Second Launch attempt once a day beginning with a launch window 90 minutes after First Launch)

Figure F-11. Launch Order Downselection and Rationale

F.4.4 Cost Sensitivity Study in "Launch Order, Launch Separation, and Loiter in the Constellation 1½-Launch Solution" [36]

The greatest level of uncertainty in the launch order analysis involves the transportation element replacement costs. Because of the uncertainty involving fixed and marginal costs for Altair and because of the uncertain nature of cost estimates in early design, large shifts could potentially occur in the cost data for all elements as the designs mature.

Figure F-12 was used to determine the optimal launch order for any set of Ares I and Ares V stack replacement costs, based on minimizing the expected cost of failure. The cost of the Ares V stack is specified on the horizontal axis, and the cost of the Ares I stack is specified on the vertical axis. The sloping red line in the center of the figure represents the break-even cost boundary. If the set of costs is below this line in the light-blue region, then the I-V launch order is preferable. If the set of costs are above the red line in the light-green region, then the V-I launch order is preferable.

The intent of the figure was to provide a visual indication of how much change could occur in the cost estimates before the launch order decision would be reconsidered. The analysis cost estimates are represented as a horizontal bar on the chart. The Ares I cost is normalized to 1. The Ares V cost is represented as a range of 1.65 to 2.88 times the cost of Ares I, which represents the full range that is produced by the possible inclusion of fixed costs. It is evident that, even at the low end of the Ares V cost range, a large margin still exists before the break-even point is reached.

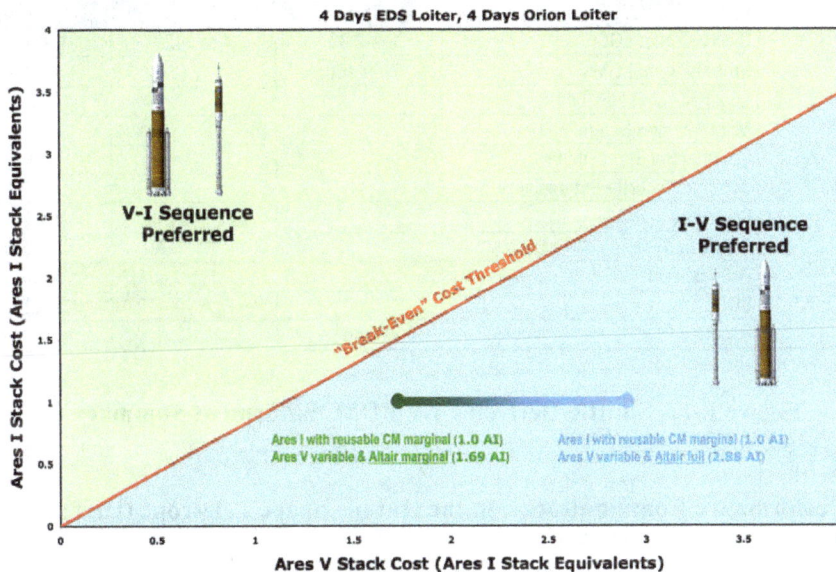

Figure F-12. Launch Decision Relative to Ares I and Ares V Stack Costs

F.4.5   Performance Communication in ESAS [20]

A summary of the ESAS FOM assessments for the Shuttle-derived CLV candidate vehicles is presented in Figure F-13. The assessment was conducted as a consensus of discipline experts and does not use weighting factors or numerical scoring, but rather a judgment of high/medium/low (green/yellow/red) factors, with high (green) being the most favorable and low (red) being the least favorable.

The Shuttle-derived options were assigned favorable (green) ratings in the preponderance of the FOMs, primarily due to the extensive use of hardware from an existing crewed launch system, the capability to use existing facilities with modest modifications, and the extensive flight and test database of critical systems—particularly the RSRB and SSME. The introduction of a new upper stage engine and a five-segment RSRB variant in LV 15 increased the DDT&E cost sufficiently to warrant an unfavorable (red) rating. The five-segment/J–2S+ CLV (LV 16) shares the DDT&E impact of the five-segment booster, but design heritage for the J–2S+ and the RSRB resulted in a more favorable risk rating.

Applicability to lunar missions was seen as favorable (green), with each Shuttle-derived CLV capable of delivering the crew to the 28.5-deg LEO exploration assembly orbit. Extensibility to commercial and DoD missions was also judged favorably (green), with the Shuttle-derived CLV providing a LEO payload capability in the same class as the current EELV heavy-lift vehicles.

	LV	Shuttle-derived CLV		
		4-Segment RSRB with 1 SSME	5-Segment RSRB with 4 LR-85s	5-Segment RSRB with 1 J–2S+
		13.1	15	16
FOMs	Probability of LOC	1 in 2,021	1 in 1429	1 in 1,918
	Probability of LOM	1 in 460	1 in 182	1 in 433
	Lunar Mission Flexibility			
	Mars Mission Extensibility			
	Commercial Extensibiity			
	National Security Extensibility			
	Cost Risk			
	Schedule Risk			
	Political Risk			
	DDT&E Cost	1.00	1.39	1.30
	Facilities Cost	1.00	1.00	1.00

Figure F-13. Shuttle-Derived CLV FOM Assessment Summary

F.4.6   Performance Communication in the Hubble Space Telescope (HST) Servicing Analyses of Alternatives (AoA) [37]

In order to communicate the cost-effectiveness of each alternative, several primary MOEs were combined into one governing metric. To develop this metric, an expected value approach was taken. Expected value theory is based on the notion that the true, realized value of an event is its inherent value times the probability that the event will occur.

The expected value approach took into account the performance of each alternative relative to post-SM4[15] capability (MOE #5), the probability of mission success (MOE #4) and the probability that the HST will have survived to be in the desired state for the mission (MOE #3), which is a function of HST system reliability and development time (MOE # 2). The calculation of expected value was the value of the alternative times the probability of the alternative successfully completing its mission:

$$\text{Expected Value} = \text{MOE \#3} * \text{MOE \#4} * \text{MOE \#5}$$

Figure F-14 illustrates the results of the combined expected value plotted against life-cycle cost. The results indicate that the disposal alternatives provided no value relative to observatory capability. The expected value calculation also indicated that rehosting both the SM4 instruments on new platforms provided higher value at equivalent cost to the robotic-servicing missions.

---

[15] Servicing Mission 4 was the HST servicing mission previously scheduled for 2005.

There was, however, a gap in science with the rehost alternatives that was not captured in this expected value calculation.

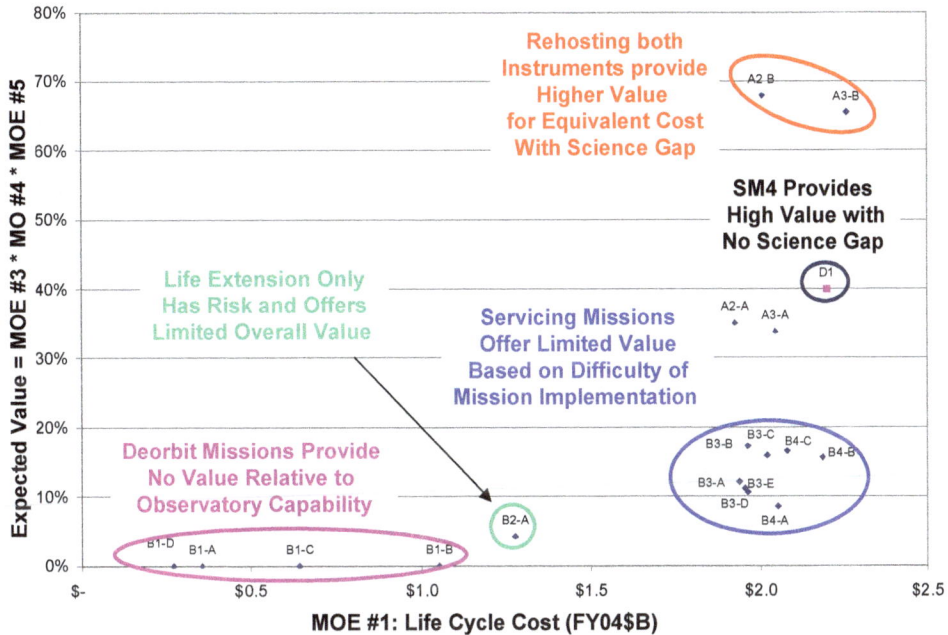

Figure F-14. Expected Value versus Life Cycle Cost

The robotic servicing alternatives cluster in the lower right corner of the plot, suggesting that the value of these alternatives was limited based on difficulty of the mission implementation, the complexity of the servicing mission, and the reliability of HST after servicing.

SM4 had costs in the same range as the rehost and robotic-servicing alternatives. It had the added benefit of higher probability of mission success than the robotic servicing missions, and did not suffer from the gap in science associated with the rehost alternatives.

www.ingramcontent.com/pod-product-compliance
Lightning Source LLC
Chambersburg PA
CBHW051222200326
41519CB00025B/7212